Valuing People in Construction

T0187548

Valuing People in Construction provides contemporary perspectives on the 'gluc' that binds the construction process together; people.

The book addresses people issues in the construction industry where behavioural outcomes impact upon business and project performance. The main proposition of the book is that as people continue to lead the completion of construction activities, their health, safety, and well-being should be seen as a priority, and valued by stakeholders. As employers and employees, the role of people in construction must be to strive for the improvement of individual lives and society. This edited collection, which is the first book to focus specifically on placing value on people in construction, focuses on people at work, gender at work, conditions at work, and respect at work. In addition to an editorial overview, the book presents tested and refined empirical work and case studies by leading construction researchers from Africa, Australia, and Europe.

Essential reading for researchers, students and professionals interested in construction management, the sociology of construction, HRM in construction, gender, work and health studies.

Fidelis Emuze is Associate Professor and Head of the Department of Built Environment, and Head of the Unit for Lean Construction and Sustainability at the Central University of Technology, Free State, South Africa. Lean construction, health, safety, and well-being and sustainability constitutes the primary research interest of Dr Emuze, who is a member of the Association of Researchers in Construction Management and the Lean Construction Institute. Dr Emuze is the Coordinator of CIB TG59 – People in Construction task group.

John Smallwood is Professor of Construction Management at the Nelson Mandela Metropolitan University, Port Elizabeth, South Africa. Both his MSc and Ph.D. (Construction Management) addressed health and safety (H&S). He is a National Research Foundation rated researcher specialising in construction related issues such as H&S, ergonomics, and health and well-being. Dr Smallwood is the immediate past Coordinator of CIB TG59 – People in Construction task group.

Spon Research

Publishes a stream of advanced books for built environment researchers and professionals from one of the world's leading publishers. The ISSN for the Spon Research programme is ISSN 1940-7653 and the ISSN for the Spon Research E-book programme is ISSN 1940-8005

Valuing People in Construction

Edited by Fidelis Emuze
and John Smallwood

LONDON AND NEW YORK

First published 2018
by Routledge
2 Park Square, Milton Park, Abingdon, Oxon OX14 4RN

and by Routledge
605 Third Avenue, New York, NY 10017

First issued in paperback 2020

Routledge is an imprint of the Taylor & Francis Group, an informa business

British Library Cataloguing-in-Publication Data
A catalogue record for this book is available from the British Library

Library of Congress Cataloging-in-Publication Data
Names: Emuze, Fidelis, editor. | Smallwood, John Julian, editor.
Title: Valuing people in construction / edited by Fidelis Emuze and John Smallwood.
Description: Abingdon, Oxon ; New York, NY : Routledge, 2018. | Series: Spon research, ISSN 1940-7653 | Includes bibliographical references and index.
Identifiers: LCCN 2017009258| ISBN 9781138208216 (hardback : alk. paper) | ISBN 9781315459936 (ebook : alk. paper)
Subjects: LCSH: Construction industry–Personnel management.
Classification: LCC HD9715.A2 V35 2018 | DDC 624.068/3–dc23
LC record available at https://lccn.loc.gov/2017009258

ISBN 13: 978-0-367-73597-5 (pbk)
ISBN 13: 978-1-138-20821-6 (hbk)

Typeset in Goudy
by Cenveo Publisher Services

Contents

Figures

Tables

About the editors and contributors

The editors

Fidelis Emuze is Associate Professor and Head of the Department of Built Environment, and Head of the Unit for Lean Construction and Sustainability at the Central University of Technology, Free State, South Africa. Lean construction, health, safety, and well-being and sustainability constitutes the primary research interest of Dr Emuze, who is a member of the Association of Researchers in Construction Management and the Lean Construction Institute. Dr Emuze is the Coordinator of CIB TG59 – People in Construction task group.

John Smallwood is Professor of Construction Management at the Nelson Mandela Metropolitan University, Port Elizabeth, South Africa. Both his MSc and Ph.D. (Construction Management) addressed health and safety (H&S). He is a National Research Foundation rated researcher specialising in construction-related issues such as H&S, ergonomics, and health and well-being. Dr Smallwood is the immediate past Coordinator of CIB TG59 – People in Construction task group.

The contributors

Abigail Powell is a Senior Research Fellow at the Centre for Social Impact, Sydney, Australia. Abigail has over 12 years' experience as an academic researcher, and her research is underpinned by her passion for social justice and equality. Abigail is an experienced project manager and mixed methods researcher with expertise in gender diversity, work-life balance, young people and social policy. Dr Powell has a keen interest in organisational governance, leadership, collaboration and workforce capacity in the social purpose sector.

Adam Rogan is a research associate at the University of New South Wales, conducting research and teaching focused on gender, sexualities, and the critical study of men and masculinities. Adam is the author of various book chapters and journal articles on the subject of gender and masculinities and recently completed a PhD in Sociology examining the relationship between young men's engagement in risky drinking and public violence and the construction of legitimate masculine identities. Adam is an experienced academic researcher with a particular focus on qualitative research methods, and previously held the position of lecturer and tutor at The University of Wollongong.

Alpana Sivam is a senior lecturer in Planning and Urban Design at the University of South Australia. Dr Sivam is an architect and a planner whose research interests lie in the interstices between these two established professions. Her current research interests include urban and regional planning, urban design, ageing, housing, and sustainability.

Andrew Dainty is a Professor in the School of Civil and Building Engineering at Loughborough University, United Kingdom. A renowned expert on the sociologies of construction practice, for the past 23 years his research has focused on the social rules and processes that affect people working as members of project teams. A parallel stream of work has developed new approaches to managing people within the construction sector, and for integrating human resource management practices with business objectives. Andrew has published over 350 papers in both academic and industry journals and is co-author/editor of nine books and research monographs. He is also co-editor-in-chief of the leading research journal *Construction Management and Economics*.

Clinton Aigbavboa is an Associate Professor at the University of Johannesburg, South Africa. His research interest is in the field of sustainable human development, with the focus on sustainable housing regeneration (urban renewal and informal housing), life cycle assessment, leadership in low-income housing, biomimicry, post occupancy evaluation and green job creation.

David Oloke is a chartered civil engineer who has extensive interest in sustainable construction research, knowledge transfer and consultancy. Dr Oloke works internationally as a consultant, lecturer, and resource person in civil/infrastructural engineering and construction management. He has also been a member of the Institution of Civil Engineers' (ICE) Health and Safety Expert Panel and the Health and Safety Register, and further serves as a reviewer for intending registrants (chartered civil engineers). His work with the ICE allowed him to be engaged as a chapter author and co-editor of the ICE *Construction Health and Safety Manual*,the first and second editions of which were published in 2010 and 2015 respectively. His research and consultancy interests now span a broad range of built environment sustainability themes and are engaging more with industrial collaborators and institutions across Europe and the rest of the world.

Elisabeth Michielsens is a Principal Lecturer at the University of Westminster. Her teaching experience focuses on research methodology in business and management. Her research concentrates on equality and diversity management in the UK, and in a comparative European context, especially in construction and STEMM-related sectors.

Erica French is Associate Professor of Management at the Queensland University of Technology, School of Business. Her research and teaching focus on women at work, including equal employment opportunity and diversity management. She is author and editor (with Strachan and Burgess) of *Managing Diversity in Australia: Theory and Practice*, and some other works with them and others individually, and together.

Fred Sherratt is a Senior Lecturer in Construction Management at Anglia Ruskin University. The research interests of Dr Sherratt focus on the championing of worker health, safety, and well-being in construction, with recent attention paid to the growing influence of corporate social responsibility in this area.

Glenda Strachan is Professor Emeritus in the Department of Employment Relations and Human Resources, Griffith University, Queensland, Australia. Her research focuses on women and work, in both a contemporary and historical setting, and in particular on the impact of national and organisational policies. She is author and editor (with French and Burgess) of *Managing Diversity in Australia: Theory and Practice*.

Helen Lingard is an RMIT University Distinguished Professor and Director of the Centre for Construction Work Health and Safety Research. On completing her PhD, Helen spent nearly ten years working as a workplace health and safety advisor to organisations in the construction, mining and telecommunications sectors. She joined RMIT in 2005. Since then Helen's research has focused on work health and safety and work-life balance in the construction industry. Her work has been funded by government agencies and industry organisations. In 2009 Helen was awarded a prestigious Australian Research Council Future Fellowship to deliver a four-year programme of research examining ways to integrate work health and safety into the design, planning, and delivery of construction projects. Helen has a strong track record in undertaking applied research in close collaboration with industry partners and in translating her research findings into industry practice. Most recently, Helen worked with Dr Michelle Turner in a participatory action research project to develop and evaluate health promotion strategies for construction workers. This work was funded by the Queensland Government under the *Healther. Happier. Workplaces* programme.

Lerato Mathebula is a construction researcher at the University of Johannesburg. Although in private quantity surveying practice, Lerato is keen on social science related research in the built environment.

Linda Clarke is Professor of European Industrial Relations at the University of Westminster, Co-director of the Centre for the Study of the Production of the Built Environment (ProBE) and on the European Institute for Construction Labour Research (CLR) board. She has extensive experience of comparative research on labour, equality and diversity, vocational education and training (VET), and wage relations in the European construction sector.

Louise Chappell is a Professor in the School of Social Sciences, University of New South Wales where she researches gender and institutions across politics, law and the corporate spheres. Louise has published widely in these areas including in her books *Gendering Government* (UBC Press 2002), *The Politics of Gender Justice at the International Criminal Court* (OUP 2016), and in many journal articles including for the *International Feminist Journal of Politics*; *Politics & Gender*; *Public Administration* and *Perspectives on Politics*.

Martin Loosemore is Professor of Construction Management at the University of New South Wales. He is a Visiting Professor at the University of Loughborough, UK and a Fellow of the Chartered Institute of Building. Martin has published extensively in the areas of corporate social responsibility, social procurement, and enterprise, innovation, risk management, productivity and strategy. Martin served as an advisor on workplace productivity and reform to the 2003 Cole Royal Commission into the Building and Construction Industry and contributed significantly to the 2009 Federal Senate Inquiry into Building and Construction Industry Improvement Amendment (Transition to Fair Work) Bill. Between 2009 and 2014 Martin also served on the Federal Government's Built Environment Industry Innovation Council and contributed significantly to the recent Productivity Commission Inquiry into Infrastructure.

Megan Blaxland is a researcher at the Social Policy Research Centre, University of New South Wales. The research interest of Dr Blaxland is in Australian and comparative family policy analysis. Her research experience has included projects on income support programmes for families; early childhood education and care, particularly family day care; and policies to support the competing work and family commitments of families.

Michelle Turner worked in industry for 20 years before joining academia. In 2008, Michelle undertook doctoral research, developing a work-life fit model for the Australian construction workforce. Dr Turner continues to focus her studies in the areas of work-life experience, resilience, and health and wellbeing in the construction industry.

Natalie Galea is a research associate and PhD candidate in the Faculty of Built Environment at the University of New South Wales, Australia. She is the lead researcher on an Australian Research Council Industry Linkage project focused on gender equity and diversity in the Australian construction sector. Natalie's research in the construction industry draws her back to her earlier working life as a construction professional and it is through this experience that she was able to be an 'insider' in the ethnographic research conducted on construction sites which is discussed in her chapter. Her PhD research draws on feminist institutionalism to highlight the role male privilege plays as a barrier to women's recruitment, progression, and retention in the construction sector.

Nicholas Chileshe is an Associate Professor of Construction and Project Management in the School of Natural and Built Environment at the University of South Australia. Dr Chileshe obtained his PhD in Construction Management from Sheffield Hallam University in 2004. His current research interests include waste management and reverse logistics.

Patrick Manu is a Senior Lecturer in Construction Project Management (CPM) at the University of the West of England, UK, where he leads and teaches on the MSc CPM programme, and he also teaches on the BSc CPM programme. Dr Manu has held academic posts in the United Kingdom and Ghana. Amongst his research interests is occupational health and safety. His research, which has been published in several outlets, has been funded by several organisations including EPSRC.

Philip McAleenan is a Researcher and OSH Management Consultant and Partner at Expert Ease International, Northern Ireland. Dr McAleenan is co-developer of the Operation Analysis and Control model for safety. His research into workplace culture, leadership and ethics reasoning led to developing the Organisation Cultural Maturity Index and to undergraduate modules on ethics thinking in construction. He publishes extensively, contributing to safety books in the United Kingdom and the United States of America (USA), and to books on workplace culture and leadership. Internationally, he regularly presents at OSH conferences, including ILO World Congresses and professional development in USA and Canada. He contributed to the International Council for Building's working commission on safety and health in construction (CIB W099) and assisted in the organisation and management of its 2015 conference, "Benefitting Workers and Society through Inherently Safe[r] Construction".

Sadasivam Karuppannan is an urban and regional planner at the University of South Australia. His research interests include urban and regional planning, ageing, housing, energy consumption, land use planning, and sustainability. Dr Karuppannan has published extensively in the area of urban planning, housing, ageing and the application of GIS in planning.

Sylvia Snijders is a Senior Lecturer at the University of Westminster. Her teaching experience focuses on organisational behaviour and research methodology in business and management. Her research concentrates on equality and diversity management in the UK, especially in STEMM-related sectors, and on best practice in the agricultural sector.

Tony Trasente has been director of Daniel Jordan Homes, one of the most prestigious construction firms in Adelaide, since 1993. He has extensive experience in design, construction and management of a variety of housing projects including detached homes, units and apartments and has received several prestigious industry awards for his projects.

Valerie Francis is an Associate Professor in Construction Management at the University of Melbourne. She is a civil engineer with a Master's degree in Project Management and a PhD which focused on gender in construction and engineering. Prior to joining academia, she was a senior structural engineer and project manager. Her research explores factors affecting organisational effectiveness, employee satisfaction and well-being.

Preface

This collection originated from the work of CIB TG59. "CIB" is the abbreviation for "International Council for Research and Innovation in Building and Construction", which is based in the Netherlands. The work of this CIB task group is evident in international conferences focused on ideas from diverse disciplines – engineering, the humanities, the natural sciences, and the social sciences. The thoughts contained in this collection, which have been disseminated on several platforms, elaborate upon the following issues: the ageing workforce, competence, disability, employment relations, gender, knowledge and skills, labour conditions, respect for people, stress, working conditions, professional life conditions, and general health, safety and well-being matters that affect people in construction (PiC). The 14 chapters in this volume take their cues from these issues, with the aim of showing how to develop, improve, challenge, and respect the people at the front line of the services and products offered in the construction industry.

Chapter 1 describes who constitutes individuals in construction (PiC), and how best to treat them. People, in the form of professionals, managers, technicians, and craftspeople, constitute PiC. The efforts of PiC produce what society requires, in most cases at a cost to them and their families. Work pressures that are increasingly exerted upon PiC in pursuit of higher productivity and profitability impact upon the quality of life of people at the front line of production in the construction industry. The centrality of PiC in terms of proficient and satisfactory outputs in the sector is indisputable. However, when bad news and empirical reports about construction health, safety and well-being are examined, the plight of people cannot be ignored. Stories of harmful exposure and consequences abound. So, Chapter 1 of the book attempts to elaborate upon who the individuals in construction are, and how these people should be treated in the industry. From an ethical standpoint, the chapter emphatically canvasses the notion of valuing people in construction, from both a business and an ethical perspective. The business case should not substitute moral reasoning when treating issues that are close to the heart of individuals that deliver our homes and every other physical immovable asset that enriches civilisation. After the introductory chapter, the edited volume is structured in three parts, which elaborate on standard construction practices, workplace diversity, and encouraged practices. I summarise the sections below.

Part I: A critical evaluation of construction practices

Construction practices, in most cases, have been the subject of emotional debates when people experience harm in the workplace. Poor handling of hazards and risks is endemic in the sector, despite evidence of the detrimental effects of such practices. Part I of this volume touches on these critical issues that affect the health, safety and well-being of PiC. The life of PiC could dramatically change for the worse if the relationship between occupational and behavioural health risks is left unattended, as illustrated by Lingard and Turner. Chapter 2 thus presents an exposition of technical and environmental issues that have an impact on the health of construction workers. Similar to other industrial sectors, poor health prevents PiC from working to their full potential. Poor health among PiC is a source of worry for their families and employers. Chapter 2 elaborates on the health of workers, which is threatened by the rigours of the physically demanding site work in the industry. The "wear and tear" from manual work is exacerbated by long working hours, which is almost a norm in the industry, job insecurity, and conflicts between labour and family life. In essence, Chapter 2 shows that PiC may experience diminished health because of the nature of construction, which produces stressors, which are usually precursors of physical injuries. As an illustration, when a construction worker is experiencing mental distress, lower back pains may result, since chronic pains in various parts of the body may be related to psychosocial risk factors. The chapter advocates for improving the health of PiC by reducing both physical and psychological health hazards in the industry.

The nature and conditions of construction site work have been the subject of many studies on housekeeping, injuries, fatalities, and the overall image of the industry. Chapter 3 of the book takes the reader through the landmines in the features of construction. The various kinds of accident causation in construction are influenced by working conditions. So, this chapter pays attention to underlying accident causal factors of construction project features (CPFs) in assessment of the degree of safety risk posed by the features. Based on an empirical study that is underpinned by accident causation models and theories, Manu confirms the known fact that early planning of project health and safety will tackle risks and reduce harm in construction. Early planning for health and safety will deploy the use of preventative measures, which will mitigate the impact of accident causation factors. The safety management literature attests to this idea, through the safety influence curve analogy. The safety influence curve, which has many downloadable variants on Google, illustrates the declining ability to influence safety positively as a project progresses towards its completion. In the current climate of a dearth of empirical work on the subject, Chapter 3 suggests that the nature of a project, the construction method used, the site characteristics, the duration of the work, and the design complexity are CPFs with accident causal potential. For example, the nature of the project will determine the nature and level of uncertainties in terms of hazards and their agents, which may transform the hazards into serious risks for a project. Putting CPFs under the spotlight is relevant because one can note agents that will act as hazard activators in construction. Manu continues the discourse on accident causal factors and project features in Chapter 4. Chapter 4 profiles the hazards and risks that should be tackled, with the appropriate preventative strategies, in the preconstruction stage of a project. I noted that in Chapter 4, CPFs that could potentially change the course of a project can be addressed by the client, the designer, the project managers, and other upstream parties to a project. Each CPF differs in potential to influence occurrence of accidents, so it is prudent that the high-risk CPFs should command the attention of project parties in the preconstruction stage of a project.

Chapter 5 introduces the reader to strategic matters affecting PiC. Unlike Chapter 3 and Chapter 4, which present operational issues that can be resolved by project parties in the upstream phases of a project, Chapter 5 focuses on a matter that is currently cause for concern in most developing and emerging countries. Indeed, the ageing workforce in the construction industry, where young new entrants are few, constitutes a serious demographic problem, which has implications for productivity in the industry. Chapter 5 highlights issues around the ageing workforce in Australia. As is the case in similarly developed countries, Australia is experiencing rapid growth in the numbers of its ageing workers. The lack of an adequate supply of workers in qualitative and quantitative terms is a major constraint, which has wider social implications in terms of development. The problem of the ageing workforce is most evident in trades that are needed to drive the construction process. Alpana and co-authors show that the Australian construction industry is facing skills shortages, which will impact upon overall performance and productivity. The revelation that the Australian construction sector has a high proportion of workers who are 45 years of age and older is concerning. The ageing workforce issue is not peculiar to Australia. Other nations are struggling to attract young school leavers into the industry as tradespeople. The situation is exacerbated by the shortage of critical craft skills in the industry. The lack of essential craft skills is a worldwide problem, and a remedy has yet to be found. So, the chapter provides a viable road map to address these issues, by identifying talent shortage gaps before employing the use of the road map as a mechanism to reduce the imbalance between supply and demand in the industry. For example, due to poor health caused by the physical demands of work in the industry, PiC tend to stop working at an early age when compared with manufacturing. There is a need to promote health in the industry, so that older workers (45 years and older) can stay longer in the construction sector. Apart from promoting health in the industry, there is also a need to retrain workers and encourage more automation, while limiting manual work in the industry.

Ageing workers worry about both work and non-work related issues. One such worry is job security, particularly where automation is becoming prevalent in a craft. As elaborated on in Chapter 2, occupational health is affected by a host of issues, one of which is job insecurity. The security of a job has implications for morale and work-life balance in the construction industry. Based on the African context, Chapter 6 presents the drivers of construction job insecurity.

The chapter explores the drivers of construction job insecurity, with the aim of minimising its detrimental influence on the work-life balance of construction workers. Aigbavboa and Mathebula show that the lack of skills, unpaid overtime, subtle racial discrimination, and gender inequality are factors that contribute significantly to job insecurity in the construction industry. These issues are not limited to South Africa. Job insecurity is a global concern, and its effects after the 2008 economic meltdown are still evident in places in Europe and elsewhere. Given the impact of job insecurity on the morale of a worker in terms of productivity and on-site safety, Chapter 6 concludes that upscaling competence, knowledge, and skills of PiC would enhance their sense of job and career security, in an industry where an ageing workforce is a reality.

Part II: A dissection of workplace diversity

Diversity in the workplace refers to the various differences between people in a company. Diversity includes race, gender, age, personality, organisational function, education, background, and more. While the construction industry also takes a comprehensive description of diversity, the male-dominated nature of the industry makes gender equity a major area in need of transformation. *Working construction: Why white working-class men put themselves—and the labor movement—in harm's way* by Kris Paap serves as an auto-ethnographic account of the life of a woman in the male-dominated construction sector. This in-depth account of a personal work experience of a woman provides insight into gender, race, class, and power diversity in the industry. The discourse on women in construction (WiC) opens Part II of this volume. In Chapter 7, Galea and co-authors advocate for a reversal in the under-representation of women, using a new theoretical and methodological lens. The chapter is relevant, as formal policies have not shifted the imbalance in gender representation in the sector. The authors use the approach of feminist institutionalism to comprehend the failure of official policies to remedy the imbalance. While feminist institutionalism provides new insights into gender equity issues, its adoption raises new methodological questions. The chapter advocates the power of alternative methodologies, such as rapid ethnographic methods, for WiC studies. From a methodological perspective, Chapter 7 has a competitive edge, as it demonstrates how rapid ethnography could produce practical insights into gender equity matters in construction. This approach is known for its rigour, as it has a theoretical foundation for the collection of field data, which is collected through close interaction with informants.

The failure of formal policies on diversity as they concern WiC is also at the centre of Chapter 8. In the chapter, Clarke, Michielsens, and Snijders confirm practice issues highlighted in Chapter 7. More importantly, Chapter 8 decries the paucity of WiC at the operational level in the industry. Anyone with a fair number of years of experience in the industry will confirm the claim that at all entry levels in the industry, participation by women is low. My work experience in two developing countries does not disprove this claim. Regardless of which measuring instrument has been used, statistics on WiC at the operational level show that participation is low, and this reality calls into question the way diversity is managed in the sector. Reported factors that are reinforcing the current situation are poor working and employment conditions, which are exacerbated by long working hours, a lack of work-life balance, and the enduring macho culture in the industry. The chapter argues that the sector should look beyond the "business case" mantra if this will produce and engender a more inclusive strategy that will activate a roadmap needed to promote a gender-neutral industry. Chapter 8 sends a clear message: to become a gender-neutral industry, the sector must embrace wider moral considerations that uphold stable employment, inclusive (and gender-neutral) education and training, and improved conditions in workplaces.

Reflections on the key points in Chapter 7 and Chapter 8 indicate that, instead of being insiders in the sector, women remain outsiders. In this regard, French and Strachan add to the debate on the state of gender equity policies in the construction industry, in Chapter 9. To address the widely held view that women are outsiders in the construction sector, Chapter 9 attempts to answer the following questions: "What are construction organisations doing to manage equity and inclusion?" and "How has management of equity and inclusion changed under varying equity legislation?" The attempt reveals that construction firms appear to be complying with equity legislation, but the extent of compliance is open to question, since a change in gender-equity imbalance has failed to manifest. In fact, compliance with the law may be undertaken through minimalist strategies that preserve the old order, namely male domination. The needed transformation, spoken about in Chapters 7 and 8, is thus mentioned again in Chapter 9. For example, where women can be found in the industry, their roles are mostly administrative in nature, owing to both entry and retention barriers, which are not helped by the masculine, dirty, and dangerous image of the industry. The chapter thus shows that women are unacceptably under-represented in the construction sector.

The final chapter of Part II takes a more inclusive approach to the diversity debate, as it expounds on a matter that affects both men and women in the industry. Chapter 10 addresses issues affecting family people in the construction sector. The notion of work-family conflict is gender-neutral, as it affects both men and women in practice. From the perspective of a civil engineering professional, Francis maps the problems associated with a lack of family support policies and a negative work-family culture. The negative role of long working hours, which is more or less the norm in the industry, is a concern that must be mitigated if work-family conflicts are to be avoided. By assessing the relationship between a supportive work-family culture and the work and life experiences of professionals, Francis shows that engineers who perceive their organisation's values to support both their work and their personal life have reported greater organisational commitment and job satisfaction and higher levels of collegial and subordinate support, as well as less intention to quit. The chapter highlights the benefits of a supportive work environment where support is strong enough to mitigate work-family conflicts and emotional exhaustion. The chapter also indicates that work patterns and work experiences of fathers and mothers are dissimilar, and as such, changes in workforce characteristics, therefore, demand a shift in management approach to promoting work-life balance among professionals in the industry. Addressing issues that are known to be detrimental to family life should be a priority for employers. Concrete interventions should be deployed to either reduce or eliminate long working hours, work inflexibility, high job demands due to productivity needs, job insecurity, and pervasive relocations of personnel without proper evaluation of the impact of such actions on family life. The contributions of this section of the book confirm the personal experience of Kris Paap. The section indicates that WiC will have to brace themselves to overcome the social, psychological, political, and physical (manual labour) battles that they face on the front line of construction.

Part III: Thinking on new practices

The imperfection of people, and the failures (errors, and violations) that continually occur in construction, always brings the competence of PiC into question. Errors and violations among PiC constitute a significant component of human contribution to the occurrence of incidents and accidents. For instance, safety-competent people should adopt safe work methods, they should detect hazards to control risks, and they should uphold general principles of prevention. Thus, in the first chapter of Part III of the book, namely Chapter 11, Oloke and McAleenan elaborate on the well-known fact that construction is a people-driven process that is prone to mishaps, owing to the nature of people with regard to errors, fatigue, and mistakes, which often lead to incidents and accidents. With "competence" as a requirement for reducing and, if possible, completely eliminating errors and violations, Chapter 11 reinforces the place of knowledge, skills, and experience in the development of a competent construction workforce. The chapter highlights the general procurement framework that has been put in place to develop a competency framework, as a precursor for developing significant health and safety competence, which will be influential in removing unsafe acts and conditions from construction sites. The chapter makes a case for competent safety clients, designers, project managers, and others.

Certainly, competence plays a crucial role in our thinking, decisions and actions. When our decisions appear to be at odds with work realities, we tend to raise questions. Such questions could suggest that there is a disjunction between our responsibilities and our accountability. Such questions will arise in situations where expert knowledge is limited or excluded from decisions that impact upon the life of PiC. A similar scenario is playing out in some construction firms, where responsibility for health, safety, and well-being (HSW) is being reassigned from the production department of the firm to the corporate social responsibility (CSR) department. While there are proponents of such reassignment of responsibility within the safety management community, Chapter 12, authored by Sherratt, questions the rationale behind such a move.

Sherratt argues that such reassignment of responsibility removes accountability for the HSW of construction workers from production practices, and instead places HSW issues under the jurisdiction of public relations. In other words, reassignment of HSW to CSR departments is negatively impacting risk assessments and methods of HSW management in practice. Informed by a case study of this phenomenon, Sherratt invites us to conceive of the "commodification of worker HSW", which is a way of illustrating the detrimental effects of such a strategic move in a firm. Reassignment of HSW to the CSR department of a firm could remove expert production knowledge and experience from critical safety decision-making processes. This could mean that people without the relevant production experience may be dictating how to handle hazards and risks on a site that they have neither worked on nor seen. Such a situation is a source of serious concern for us all.

The recorded errors and violations in construction, which constitutes a complex sociotechnical system, have been a source of concern and the object of interventions for decades. The fact that novices in actual on-site work could be deeply involved in the design and implementation of safety management systems (SMSes) would not set at ease PiC, who are on the front line of production in the sector. Poorly designed and implemented SMSes, which sometimes amount to mere publicity stunts or slogans, are the main focus of Chapter 13. Emuze and Sherratt suggest that zero harm as an SMS is now increasingly used across many industries, and the construction sector has bought into it. Similar to other ideas, such as that of lean production, which originated in the manufacturing sector, adaptation of the zero harm vision to the construction industry is a subject of debate in the community of practice. There is criticism and scepticism concerning the way the term "zero" is understood in safety management. Drawing on case studies and research concerning zero harm in practice, Chapter 13 explains how zero harm can work for the construction industry. Use of zero harm as a philosophy, rather than as an absolute numeric target, supported by significant change in existing processes and practices that are proving ineffective, is essential for success. The chapter argues that the underpinning philosophy for zero harm should drive a wider positive safety culture, which will enable collaborative development of such change in practice. The authors of the chapter advocate for a shared meaning of the vision of zero harm in the industry. A shared meaning of this vision should tackle the harm that currently affects people in construction who are working on sites, in organisations, and along supply chains.

As mentioned in Chapter 1 and Chapter 8, a wider ethical reasoning is now required in the industry. The question is "Which ethic would serve as a point of departure from the norm?" Chapter 14 brings the work of Immanuel Kant to life, by applying his categorical imperatives to construction. The Kantian notion of respect for persons (RfP) serves as a philosophical concept for ethical or moral consideration. Chapter 14 builds on the insights in chapters 12 and 13, which run counter to the spirit of the notion of RfP, in that people are seen as ends in themselves (and not as a means to an end), and the need to make a drastic change in mindset. The concluding chapter of the book adopts a philosophical approach to the issue of respect or disrespect for workers in the industry. Chapter 14 advocates for use of the original Kantian principle of RfP in the construction sector. Emuze and Smallwood explicate the notion of RfP, and its implications for ethical thinking and practice in the construction industry. A shift away from the "business case" and other utility principles in engagement of workers in the sector is paramount for addressing many of the problems in the industry. The chapter suggests that the Kantian notion of RfP is a way of conceiving of profit as the outcome of good construction practices, rather than as a goal.

The eclectic nature of this edited book is a plus for anyone keen on expanding their knowledge on a wide array of ideas that can be explored to improve the lot of PiC. I thank all the contributors for the breadth and depth of ideas shared in the book. I must confess that this edited work was borne out of involvement in people-related research and practice activities by John Smallwood and me. John Smallwood, the co-editor of the book, who stepped down from the role of International Coordinator of CIB TG59 in 2016, identified a need to compile such a volume. I thank him sincerely for providing the required motivation to produce this book. He also persuaded me to succeed him as International Coordinator of the group in 2016. Apart from the encouragement of John Smallwood, who has become a father figure to me, I also want to thank Alfred Ngowi and Mohamed Mostafa of Central University of Technology, Free State, for supporting this project with seed funding, which allowed the work to be started and completed as planned.

I am also indebted to the willingly given expertise of colleagues who gave of their time to methodically review the chapters in the book. Their criticisms and suggestions have without doubt enhanced the scholarship and knowledge contributed to the book. I am grateful to Ciaran McAleenan (Ulster University), Philip McAleenan (Expert Ease International), Radhlinah Aulin (Lund University), Tony Putsman (Xenophon Project Services Limited), M. Reza Hosseini (Deakin University), Paul Bowen (University of Cape Town), Emmanuel Aboagye-Nimo (University of Brighton), Jane English (University of Cape Town), Tessa Wright (Queen Mary University of London), Beverley Lloyd Walker (RMIT University), Emmanuel Manu (Nottingham Trent University), Nuhu Braimah (Brunel University London), Claire Deacon (Occumed CC), Mike Behm (East Carolina University), Kenneth Yiu (University of Auckland), Simon Smith (University of Edinburgh), Evelyn Teo (National University of Singapore), Bo Terje Kalsaas (University of Agder), and Mohamed Sherif (Griffith University). The ideas in this book have been inspired by individual chapter contributors, and these ideas have been refined by the mentioned reviewers, although editorial biases in this book are, of course, my responsibility. I am very grateful to the chapter contributors, who had to collaborate in the face of various different work demands and deadlines, to produce each chapter. For English-language editing of each chapter, I have Anthony Sparg and Renée van der Merwe to thank, and at Taylor and Francis in London, I am grateful to Matthew Turpie and Ed Needle for the freely given professional support.

Without denying the obvious, I am always indebted to the generosity and thoughtfulness of my wife and our boys, Imole and Irawo, for allowing me to steal from family time to fulfil scholarly responsibilities and RCCG (Redeemed Christian Church of God) church obligations, without a complaint. While I am mindful of always trying to strike a balance between work and family, sacrifices are made for the greater good. A splendid example of my wife's kindness is the fact that instead of celebrating our ninth wedding anniversary with her today, I am away from her, teaching government construction contractors, and also writing this foreword in between breaks in Parys, South Africa. Indeed, our family constitutes the sum of who we are – with them we can achieve the impossible.

Fidelis Emuze
Bloemfontein, South Africa
17 January 2017

Part I

A critical evaluation of construction practices

1 People in construction

Who they are, why we need them, and how to treat them

Fidelis Emuze

All our knowledge begins with the senses, proceeds then to the understanding, and ends with reason. There is nothing higher than reason

(Immanuel Kant)

Who are the people in construction?

To answer the question 'Who are the people in construction', it is necessary to describe what construction is from the site operational angle. It is a known fact that construction makes use of a range of labour, material and equipment resources to realise a built physical asset, which could be a building, an airport or a dam. The organisation and use of required resources can be viewed at both strategic and operational levels (Halpin & Riggs, 1992). At the operational level, decisions and actions relating to the duration of the project and the use of people and machines must be considered. Physical or manual work involving the use of concrete, steel, glass, plastics and an infinite number of items must be undertaken to construct and finish a project, either directly by people or by machines directed by individuals. It is the production level in construction where people act as the 'dynamic' resource that either pulls or pushes input needed in operation. The output level in construction is where planning, design, execution, analysis and control measures work together to realise a facility (Halpin & Riggs, 1992; Koskela, 1997). The actors at the production level of construction can thus be located in the three broad categories illustrated in Table 1.1.

People are used in the realisation of physical production activities in construction. Craft workers are the individuals involved in manual work through various craft activities, and they outnumber technicians, professionals, and managers on a typical site and the industry as a whole. Professionals, managers, technicians, and artisans collectively undertake the transformation, flow and value aspects of production in construction (Arbulu, Tommelein, Walsh & Hershauer, 2003). This book, therefore, pays close attention to 'boots on the ground' managers, professionals, and workers as the people in construction (PiC) that every stakeholder should endeavour to value. Instead of only prioritising health and

Table 1.1 Broad categories of people in construction operations

Category	Description
Craft	Craft occupations carry out manual/physical work on construction sites. Roles include those of bricklayer, general worker, painter, roofer, scaffolder, plant operator, site joiner, and the like. Regarding entry, people enter the sector through formal or informal apprenticeship. Self-employment is, therefore, common among tradespeople. It is notable that a significant portion of people in the craft are non-permanent employees who often work under the direction of subcontractors or contractors who determine the working hours and the pay rate. Artisans can work as tradesmen and labour-only subcontractors. The majority of the general operatives also work in the informal sector where they are unrecorded in official statistics. In some parts of the world, migrant workers are located in the informal sector. Workers in the informal economy are not covered by social protection, labour legislation or other protective measures. A notable feature of the informal sector is the widely reported low and irregular pay that is underpinned by volatile hiring. Precarious work engagement is also found in these trades. Dangerous work involves part-time employment without the necessary social and legal protections. Craft workers often make up the largest portion of people found on a construction site. In general, artisans are involved in collective agreements arranged by their unions.
Technical	Technical roles in the construction industry support the work of professionals and managers. These roles have to use practical knowledge on sites without actually doing the manual work. Technical jobs on sites include those of the building technician, roofing technician, plant technician, site engineer, and the like. Vocational education and training that is supported by work experience are usually required for entry. Technicians could work to progress to supervisory and managerial roles in the industry. Their working hours are regular through the months. Although it is not common, part-time employment is also used to engage technicians. They also work for contractors and subcontractors (to a lesser extent) who apply statutory provisions and protective measures. Among technicians, job security is relatively stable, and their pay depends on abilities, seniority, and qualifications. Technicians may also benefit from provisions of collective agreements. However, in some countries, technicians are not included in collective agreements.
Professional / Managerial	Professional and managerial workers in the sector oversee the spectrum of all construction activities, from conception to commission, through design, procurement, planning, and management. At the top, there are senior executives and other business-related roles. At the middle and site operational levels, there are diverse functions that are not limited to those of civil engineer, construction manager, project manager or quantity surveyor, among others. These workers are employed to do regular work in the sector with a measure of job security and compensation benefits. They tend to be fewer than technicians on a regular project, and they also benefit from collective agreements.

Source: Several authors such as Langford and Agapiou (2007), for illustration purposes only

safety (H&S), valuing people is desirable since what someone values is not subjected to changes that are synonymous with priorities. In equivalent terms, most 'boots on the ground' people fall under the category of blue-collar workers on construction sites. The administrative professionals are involved in site management while the blue-collar workers are the people who undertake the physical work on project sites. The competencies and abilities of people determine the nature of construction operations and eventual project outcomes. Education, training, experience and competence are required pre-requisites for entry and movement within the three categories shown in Table 1.1. For instance, when competence is in place, treating people right would mean open entry doors that allow the progression of craft workers to technical and professional or managerial positions in the industry.

The pressure to optimise the performance of people while balancing the achievement of cost, environment, health, safety, quality and productivity objectives is a source of tension in practice. For example, while getting the job done as quickly and cheaply as possible may be pushing practices on site, the real focus of contractors would seem to be on maintaining an appearance of safety (Paap, 2006). Thus, work pressure has been increasing over the last decade, based on the reported low productivity in the construction industry where research and development (R&D) are lagging behind the money spent in manufacturing, aerospace and other industries (Forbes & Ahmed, 2010). While the need for productivity improvements is not disputed and could be justified with various forms of information, the issues that impact upon the health, safety and well-being of PiC cannot be overlooked. People should enjoy their working conditions in the industry while other stakeholders should be in agreement about keeping them alive, healthy and safe – at all costs.

Concerns around PiC have gained traction in construction management research and practice in recent years. Evidence from three books that have addressed people within the context of what they do in the construction industry suggests that the contributions of individuals in the sector cannot be rivalled. In *People and Organisational Management in Construction*, Naoum emphasises the fact that the progress of a construction firm is dependent on the clarity of operations, the quality of its people, the availability of the required resources and the appropriateness of the structure and management systems used by the organisation (Naoum, 2011). Through organisational and human behaviour concepts, this book shows the centrality of people regarding their managerial and social functions that determine effectiveness. The book focuses on the strategic or business aspect of construction management, which provides a systemic approach to what people do at work regarding their interactions that decide the success or failure of an organisation.

In the edited book, *Human Resource Management in Construction: Critical Perspectives*, Dainty and Loosemore (2013) continue the emphasis on the need to appreciate people more. The critical perspectives shared in this book targets a crucial strategic issue in construction – human resource management (HRM). The HRM in the book mentioned above brings to the fore the lived experiences and realities of people working in the industry (Dainty & Loosemore, 2013). By so doing, it outlines many issues to be addressed concerning the nature of construction HRM. The skewness of employment and HRM practice in the industry in favour of outsourcing and self-employment with its attendant complexity are well articulated (Dainty & Loosemore, 2013). Among other calls for action, this book questions practices that claim to improve performance and productivity (and inadvertently do so at the peril of people). While recognising the usefulness of current construction HRM practices, this book draws attention to the need to evolve new thinking about people management in the construction industry. (Box 1 provides a critique of people management.)

In an earlier book co-edited by Dainty, Green, and Bagilhole (2007), critical discourses on the employment of people and the culture in the construction industry were highlighted. The book, *People and Culture in Construction*, acknowledges the people-intensive nature of the project-based industry where the impact of competitive tendering, fragmentation, the large number of small firms, the limited number of large businesses, cyclical demand, fluctuation in market, structural flexibility, cultural diversity, different employment regimes, fixed location of products, male-dominated workforce, temporary work teams and mobility of labour affect everyone – internal and external (Dainty, Green, & Bagilhole, 2007). Several issues that have remained pervasive in the industry have been addressed, and the contributors to the book advocate the following (Dainty et al., 2007).

- A better understanding of industry employment practices and their corresponding impact on workers is required.
- A way to prevent employers from running away from their skill development responsibilities in the industry has to be evolved.
- A paradigm shift from a traditional attention on managers at the expense of the operational (productive) workforce has to be encouraged.
- A focus on people issues, as opposed to performance outcomes, is crucial to the continued relevance of the industry.

Box 1.1 Critique of people management

It is advocated that it would be good to stop thinking about managing people. Rather, thinking should shift to how, with respect, to trust all workers in the industry to handle the process and production of construction products. Managing people tends to dehumanise workers by ignoring their ability to think before they act critically. It also ignores employees' contributions to the success of each task and their individual life in general. Thus, the new thinking applies to craft, technical, professional, and managerial workers in the industry.

Nine years after these issues were highlighted by Dainty et al. (2007), significant positive steps are yet to be extensively reported – not much has changed. For example, the 2008 economic meltdown has more or less eroded any perception of permanent job security in the industry, regardless of the category of employment. As opposed to embracing skill development investment and interventions that would benefit the interest of people in the industry, employers are increasingly seen making decisions to either preserve their small margin in the industry or to keep their companies in business. The points argued in the third and fourth bulleted points above align with the chapters of this particular edited work that highlight the need to involve all categories of workers in the quest for improved construction industry experience that is backed up with enhanced job performance, whole-life well-being and finished products. This particular volume places more emphasis on understanding the perspective of construction operatives and it is hoped that such comprehension of people issues will inform practice.

As mentioned, the skewed focus on the performance outcome of projects at the expense of the needs of the people that do the work should be discarded as such a narrow focus has consistently shaped and rebranded the problems in the industry. These pervasive problems include cost overrun, rework, defects, poor employee morale, delays, conflicts and in-fighting, fatalities, permanent injuries, chronic diseases, and many more quagmires that have engendered the wrong image of the industry (Abdelhamid & Everett, 2000; Adsul et al., 2011; Chi et al., 2012; Dainty et al., 2007; Dekker, 2014; Edwards & Nicholas, 2002; Glendon & Litherland, 2001; Respect for People Working Group, 2004; Kuruvila et al, 2006; Loushine et al., 2006; McGeorge & Zou, 2013; Murray et al., 2001; Ness, 2010; Ness & Green, 2008). In effect, a focus on people issues in construction should begin with the rationale for such attention as briefly explained in the next section of this chapter.

Why do people need attention in construction?

There are many reasons why people require attention in construction because projects bring capital and people together to produce the finished facility. The capacity to create constructed assets lies with these resources. The number of individuals involved varies from project to project. The level of involvement of people in a construction process also makes a difference. In practice, the variation in the number of people engaged is determined by the type, size of a project, and the project stages. Because of the constantly changing nature of work (due to inclement weather and demands of a different stage of work) employers have the leverage to let workers go without providing a reason beyond 'work is slow'. For workers who are keen to remain working for extended periods, securing the goodwill of employers is another layer of pressure to be surmounted. Such a worker has to prove to be good, worthy, and cost efficient. Beyond the number and scale narratives, reports have argued that improvements in how PiC are managed would have knock-on effects on coordination, cost, environment,

H&S, planning, productivity, quality and other performance targets in the industry (Egan, 1998; Latham, 1994). As illustrated in Box 1.1, how people are
managed in the industry has nonetheless remained a grey area (Dainty et al.,
2007). From the 'people issues' angle, PiC need attention because of the nature
of the working conditions in which their contributions to the industry and society are mostly enacted. For example, gone are the days when "…workers were
neither allowed to call the day nor, in general, able to avoid working in the rain"
(Paap, 2006: 32). The rules governing working conditions on construction sites
are increasingly reported to be marginalising workers because of productivity,
timeline, and cost concerns.

Construction work influences about the likelihood of harm can be used to
illustrate a point here. There is a need to pay attention to people who have their
boots on the ground on construction sites. Construction work is undertaken in
an industry that is structured and cultured in ways unfavourable to the H&S and
well-being of its people. For example, various forms of issues lead to stress, which
in turn produces hazards and risks, which produce either safety impacts or health
impacts. The interaction of people with hazards and risks thus creates accidents
and illnesses as witnessed in the industry. Invariably, this illustration alludes to
the widely reported realisation that influences around construction work have
made hazard, risks, stressors, accidents, illnesses and harm systemic in the sector.
The influences show how each element in the system balances and reinforces the
others in conformance with the work of Senge (2006). It is, however, important
to note that mitigated harm is often not an inevitable result of efforts to eliminate hazards in the industry. While not an exhaustive list, stressors that create
agents for hazards and risks in construction work include long working hours
made attractive with overtime incentives, job insecurity, neglect of diversity in
the workplace, reduced professional worth, cost and time pressures, and conflicts
within temporary work teams (Sang et al., 2007).

The impact of harm in the industry is monumental despite several attempts to
compute the direct costs as well as the indirect costs of accidents, apart from the
money spent on compensation insurance claims. The world over, a harm that is
evident in diseases, injuries and fatalities is endemic in the construction industry
despite increasing evidence that the situation could be different. This perception
only emerges from countries where statistics are available for people in the
formal sector of the industry. Migrant workers in some instances are not captured
in official statistics, yet they tend to be harmed while making contributions to
society through construction work. From diseases to fatalities, migrant construction workers are vulnerable in this industry where people issues play second
fiddle to cost and productivity targets in practice (Adsul et al., 2011; Bust et al.,
2008; Ghaemi, 2006; Kuruvila et al., 2006; Trajkovski & Loosemore, 2006).

Although no-one signs up to encounter harm, Table 1.2 and Table 1.3 are
examples of what people could face while working in construction. Among the
hazards and risks sampled in Table 1.2, falls from height (from exposed platforms,
scaffolds, roofs, and the like), being caught in between objects (for example,
machinery for joinery), total collapse (from buildings under construction or in

use, and other structures such as excavations), being struck by objects (mobile objects such as vehicles), and electrocution (through contact with naked power lines) constitute the leading accident hazards and risks in the construction industry (Carter & Smith, 2006; Jeong, 1998; Kisner & Fosbroke, 1994; Pinto et al., 2011; Toole, 2002). Construction tasks are associated with hazards and risks owing to electrical, mechanical, physical, biological, chemical and radiation agencies (see Box 1.2 on the conflation of hazard, risk, and loss).

Valuing people is, therefore, a required attribute that should pervade all strata of the construction industry if the hazards and risks sampled in Table 1.2 (and much more) are to be eliminated. Some of the events in Table 1.2 could be consequences in their own right in certain situations. Moreover, Table 1.3 indicates a range of consequences that follow the unmindful treatment of hazards and risks in construction. The table shows the various harmful effects of poor H&S in the industry. In the worst cases, fatalities are recorded in the industry through onsite and offsite accidents (electrocutions, falls, cranes and cave-ins, among others) and non-fatal diseases such as any one of the common musculo-skeletal disorders (MSDs) (Beavers et al., 2006; Chi et al., 2009; Dong & Platner, 2004; Janicak, 2008; Jeong, 1998; Kisner & Fosbroke, 1994; Ringen et al., 1995). Recently, the issues around well-being in the form of poor work-life balance, burnout, fatigue, lack of social support, and depression have come to the fore as critical problems that accompany individual work regimes in construction (Lingard et al., 2007; Lingard & Francis, 2004; Sang et al., 2007; Watts, 2009).

The leading illness hazards and risks in the construction industry have been found to be back injuries from carrying heavy loads and working in awkward

Box 1.2 Conflation of hazard, risk, and loss

To conflate hazard and risk to form a whole narrative about a situation in a workplace prevents a complete differentiation of the harmless and harmful states of sources or agents. The conflation thus makes it difficult to realise benign states. Mindfulness regarding what is a hazard, risk, and loss is required to correctly understand what constitutes health, safety, and well-being at work. A hazard is a source of potential harm that is harmless until it is activated to become a risk, which could lead to a loss (a negative outcome such as injuries and fatalities). A risk or threat could also be changed into a hazard when it is successfully manipulated into a harmless state. Improper analysis, loose language, and superficial interchanging of concepts lead to such conflation that often ignores the transition between hazard and risk. In brief, a hazard is a source of risk or threat in a harmless state and risk has the same origin in a state of harm. Hazard and risk are different states of the same agent or source. In other words, a hazard is a potential risk that is contained and dormant. Both hazard and risk have the ability to cause a loss.

Table 1.2 Examples of exposure to hazards and risks in construction operations

Random set of hazards and risks found on the site of a typical construction project

Hazard

Falls from height	Falls at same level	Fall of objects	Struck by objects
Stuck between objects	Running over/trips	Collapse/cave in	Collisions – vehicles
Collision – equipment	Explosions	Slippery surface	Pointed objects
Cutting objects	Irregular floor	Shock – fixed objects	Vibrations
Noise – exposure	Hits on the head	Sting	Friction
Machine – exposure	Lifting – uncontrolled	Pulling – uncontrolled	Pushing – uncontrolled
Poor handling – tools	Equipment – loose	Plant – loose	Movement – repetitive
Working – poor posture	Working – untrained	Working – long hours	Working – alone
Electrostatic discharges	Electric shocks	Hold – rotary parts	Visibility – low
Humidity	Confined space	Site layout – wrong	Fragments – loose
Metal – hot	Metal – projections	Exposed nails – pinch	Exposed nails – cuts

Exposure & Risks

Intoxication – alcohol	Intoxication – drugs	Heat & cold	Asbestos
Flame	Fire	Radiation – ultraviolet	Radiation – infra red
Radiation – solar	Radiation – ionizing	Radiation – laser	Fumes & dust
Acid	Gases	Solvents	Toxic products

All hazards and risks have to be either eliminated or isolated to remove the likelihood of harm

Source: Several authors such as Dias (2009) for illustration purposes only

Table 1.3 Sampled forms of harm suffered by people in construction

Examples of health, safety and well-being issues in construction

Accident/Injury

Fatalities	Back injuries	Loss of bodily functions
Burns	Hearing losses	Amputations of body parts
Lacerations	Fractured and broken bone	Vision impairments

Illness

Respiratory diseases	Neurological diseases	Communicable diseases
Dermatological diseases	Skin diseases	Musculoskeletal diseases
Heart diseases	Cancers	Poisoning

Stressor

Poor work-life balance	Inadequate social support	Low job satisfaction
Poor relationships	Inadequate family time	Exhaustion / burnout
Depression	Poor emotional intelligence	Fatigue/weariness

All forms of harm are recursive regarding their impact on people in construction and their networks

Source: Several authors such as Sang et al. (2007), for illustration purposes only

positions; respiratory diseases from inhalation of dust, fumes and other toxins; MSDs (such as sprains and strains of the muscles); hearing loss from extended exposure to noise by operatives (often earthmoving workers); and skin diseases (Dias, 2009). As a minimum prerequisite, the application of the general principles of prevention (GPP) on every project should be mandatory in construction. Avoiding risks, evaluating risks that cannot be avoided, combating all risks at source, adapting every task to people, adapting to technical progress, replacing dangerous materials and processes with non-dangerous and safer alternatives, developing a detailed GPP policy, placing high priorities on collective protective measures (as opposed to individual protective measures), and giving proper instructions to workers all constitute principles that should be applied, both in the design and implementation phases of construction projects (Aires, Gámez & Gibb, 2010; Gervais, 2003; Holt, 2008; Ling, Liu & Woo, 2009).

It is clear that people deserve to be treated properly to avoid the problems illustrated in Table 1.2 and Table 1.3. The next section of the chapter is a short précis of the minimum treatment that people merit in the industry.

How to treat people in construction?

The notions of respect for persons (RfP) and dignity have received sporadic attention in the construction management literature. RfP in construction has been debated by researchers and practitioners, whereas dignity has received less attention. These concepts are central to ethical discussions and the moral law with which most players in the industry concur. These two concepts are central to how to treat people in construction, irrespective of their rank and level in the industry. Their importance is premised on the rationale that basic dignity and RfP should be common to all human beings (Please see Chapter 14 for more explanation on the original Kantian notion of RfP). To be succinct, Figure 1.2 is suggestive of what people in construction merit. RfP that is widely recognised in the industry should be underpinned by human dignity (*Menschenwurde*). The German word, *Menschenwurde*, points to the dignity that all people have – the human value. Human dignity provides the basis for human rights (Andorno, 2009; Cranor, 1983; Darwall, 1977; De George, 2011; Donnelly, 2013), which everyone should enjoy since people are born free and equal in dignity and human rights (Glendon, 1997). A question to be asked is to what extent this declaration by the UN is ingrained in how people are treated in construction, especially the non-managerial cohort in the industry. Respect for human rights is inclusive of RfP, which has been used to examine ethical issues regarding labour relations, HRM practices, pay equity, employee safety and human rights at the workplace, to mention but a few.

The notion of human dignity suggests that some people should not be subjected to less respect than their contemporaries. Human dignity goes against treating someone as a means only, instead of an end. Incontrovertibly, the plight of migrant construction workers in particular regions of the world is not dignified in all ramifications (Adsul et al., 2011; Bust et al., 2008). This migrant worker reality in the construction industry is against the notion of universal human dignity that pertains to all people without the possibility of being lost as long as a person is alive. The living condition of migrant workers in Qatar as a result of the 2022 World Cup construction projects is a vivid reminder of the inhuman treatment of people in the industry. In essence, the notion of human dignity refers to the idea that every individual is inherently worthy, and therefore each merits respect and great consideration (Donnelly, 2013; Lee & George, 2008; Taylor, 2003). In other words, a person such as a migrant construction worker should never be treated as a thing or a resource for gain. (Please see Box 1.3 on dialogical contradiction between human right and dignity.) The idea of human dignity and RfP espoused in this book and specifically in the concluding chapter are at variance with the use of 'business case' arguments for encouraging sound people management practices. The business case arguments are targeting effectiveness that would boost economic performance instead of having amplified humanity as the ultimate aim. In essence, correctness in dealing with people is not seen as treasured in RfP construction reports, but only regarding its contributions to profit making (Melé, 2014).

Apart from universal human dignity, the concept of human quality treatment (HQT) has aspects that are essential for upholding the proper treatment of people in construction. The five levels of HQT include maltreatment, indifference, justice, care and development (Melé, 2014). The first two levels of HQT (abuse and indifference) are unsuitable for people, regardless of the sector where they are employed. For instance, 'indifference' could mean the disrespectful treatment of individuals by not acknowledging their personhood. In contrast to maltreatment and indifference, Figure 1.1 shows that justice, authentic care, and development (or advancement) would enhance the notion of RfP in the workplace. Quality as a concept is not new in the construction industry lexicon. Standard terms related to quality include quality assurance, quality control, quality management systems, and total quality management. The word 'quality' and the mentioned terms all refer to the need to ensure excellence in service, process, and product. Human quality is about people who share similar ideas of

Box 1.3 Dialogical paradox of human right and dignity

Respect for people, and by extension, the dignity of people goes beyond human rights, which is evidently not in existence in wage labour offered to a large section of workers in the construction industry. Wage labour separates the workers from themselves and the world in which they find themselves. Human rights, therefore, exist because these rights are lacking in wage labours and the treatment of people in various industrial sectors of the global economy. For example, migrant workers should have access to decent jobs and obtain fairness in remuneration, for wage labour to be eliminated.

excellence in an organisation. Thus, HQT is about dealing with people in a way appropriate to human conditions (Melé, 2014). The treatment entails acting with respect for the human dignity of people and their rights and caring for their problems and legitimate interests. As opposed to the concepts of utilitarianism and contractarianism, HQT is aligned with Kantianism as explained in Chapter 14 (Johnson, 1974; Kant, 2012; Kant & Friedman, 2004).

For readers who are not familiar with utilitarianism, it is a theory that provides a rationalistic tool that carries the risk of a lack of respect for human rights if the financial calculations do not favour such a line of action. Although contractarianism has elements of human quality, it is strictly focused on the letters of contracts and agreements. These two philosophical concepts are not completely in agreement with the Kantian version of RfP. But aspects of HQT have similarities with various categories of RfP. HQT has five distinct organisational levels (Melé, 2014):

- Level 1 is about maltreatment that is offensively noticeable. Injustice at this level is perpetrated through the abuse of power and exploitation. The abuse of people through exploitation, aggression, ill-treatment, aggressive treatment, psychological (bullying, threats, fear inducement) and sexual harassment leads to injustice. Manifest injustice occurs with the manipulation of people through underhand tactics, deceit, lies, and unfair discrimination.
- Level 2 refers to indifference, which entails the lack of proper acknowledgment of personhood. The lack of recognition then eventuates in disrespectful treatments. At this level, people are treated with limited or no personal appreciation, and they are used as little resources without paying attention to their humanity. For example, managers will fail to discuss whether a decision would either help or harm the people that would be affected. Thus, at this level, people are treated as means to economic ends.
- Level 3 is about justice that people experience regarding the respect for individuals and their associated rights. The dignity of personhood is respected and their rights are not compromised. The justice at this level entails acting with integrity, fairness in pay, and fulfilment of agreements.
- Level 4 is about authentic care, which is observable in genuine concerns for the legitimate interests of people. The care allows the provision of needed support for people in resolving their problems. At this level, people are recognised and respected for their rights with particular care regarding all their concerns and interests. For example, support through empathy, compassion, and emotional intelligence is offered to people who are battling with personal problems that would affect their well-being. Such problems could be poor health or family conflicts.
- Level 5 is about development, which encompasses human flourishing, mutual esteem, and reciprocity based on friendship. At this level, the development includes ensuring that people have justice, resolving the problems of people, paying attention to care requirements, and acting with a sense of service that is boosted by active friendliness.

Without gainsaying, three HQT levels among the five apply to how to treat PiC. These include justice, authentic care, and development. As an illustration, someone of 'good' who is respected, will have justice, genuine care (care that is not subject to emotional swings), and development as significant experiential aspects of such respect. Figure 1.2 provides more details on these levels. Justice in Figure 1.1 refers to respect towards persons and their rights; authentic care is the concern for the legitimate interests of individuals and the need to support them regarding solving their problems, and development in the figure refers to the willingness to work for the mutual esteem and prosperity of people.

According to Figure 1.2, indifference in the treatment of people should be eliminated because it is a complete lack of recognition of the personhood of an individual – the humanity in every human being. Indifference is contrary to human dignity and right, and so it is at variance with the worthiness of each person. The literature shows that indifferent treatment of people is found in working teams and is quite common in small and medium-sized enterprises (SMEs) (Teo & Loosemore, 2001). Given that SMEs and temporary work teams are dominant in the construction industry, indifferent treatment of people is a problem in the industry. For example, indifference from managers towards working conditions on a project site could be linked to many housekeeping-related hazards and injuries in construction (Haslam et al., 2005; Lipscomb, Glazner, Bondy, Guarini & Lezotte, 2006). Even hypocritical behaviour from managers

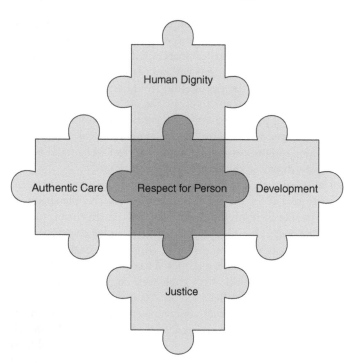

Figure 1.1 An illustrated model of how to treat people in construction

		DO YOU HAVE IT?	
		NO	YES
D O Y O U W A N T I T ?	**Y E S**	*Achieve Authentic Care*	*Preserve Development*
		✓ Show concern for the interest of people ✓ Be sensible and help people to resolve problems ✓ Consider individual vulnerability in decision making ✓ Strive to improve working conditions ✓ Tolerate faults and errors with conditions ✓ Reconcile work with family life ✓ Maintain conciliatory attitude towards interpersonal conflicts	✓ Act with a real respect for collaborators ✓ Favor integrity and moral behaviour within the firm ✓ Act with a sense of service and cooperation ✓ Foster collective initiative, creativity, participation, sense of responsibility and commitment, and mutual respect ✓ Be aware of the talents of people and help them to enhance it ✓ Foster the acquisition of professional know-how
	N O	*Avoid focus on Justice only*	*Eliminate Indifference*
		✓ Recognise and respect human dignity and rights ✓ Fulfil all agreed pacts and contracts ✓ Ensure fairness in remuneration, performance appraisal and every other managerial system ✓ Ensure fairness in employment and dismissal ✓ Ensure transparency and truthfulness in communication ✓ Act with good faith and loyalty ✓ Ensure equity in distributions ✓ Respect confidentiality ✓ Be willing to resolve and change unfair situation	✓ Dealing with people with limited or no appreciations ✓ Not paying attention to empathy ✓ Evaluating decisions only from legal and business views ✓ Showing considerations,only for the sake of profit ✓ Speaking and acting without respect and courtesy ✓ Acting with unfounded suspicion regarding the good will of co-workers ✓ Not listening to co-workers or ignoring the experience of co-workers ✓ Not appreciating workers when it is due ✓ Acting with indifference regarding the spirituality of workers ✓ Indifferent to role and reporting line demarcations ✓ Strict compliance to the law without due regard to the peculiar situation of affected people

Figure 1.2 Sampled generic issues for promoting high-quality treatment of people (adapted from Melé, 2014: 467)

and colleagues in work teams that is aimed at profit making alone at the expense of people issues should be not be condoned on sites. The attributes shown in the justice quadrant of Figure 1.2 should be seen as a starting point in how people are treated in construction. Justice can be present in many construction companies, especially the large firms that are committed to abiding with UN charters on human rights wherever they work. The issue is that without government enforcement, justice may not be widespread in the sector. Justice in the HQT framework includes honouring contracts, operating with integrity, ensuring fairness in remuneration and having the utmost respect for the truth.

It is herein argued that the components in the bottom right quadrant of Figure 1.2 should be avoided in every enterprise. A 'No' should follow a 'Do you want it?' question regarding these components. Although the literature may allude to the fact that most enterprises operate within the 'justice' component, it is important to state that the component should be a starting point in how people should be treated. In contrast, the elements at the top quadrants of Figure 1.2 are real expectations of how people deserve to be treated – a 'Yes' should be the response to the question, 'Do you want it?' Apart from wanting these components, efforts should be made to have them in every construction organisation. This idea implies that when 'No' is the response to 'Do you have it?', efforts should be made to achieve the component. For example, authentic care involves real concern for people regarding their needs and legitimate interests. This care is inclusive of empathy and all elements of human dignity and rights. Building on authentic care, people should work where their developmental progress is assured. Although the elements of this component are not common in many firms, authentic care and development are worth implementing in construction. The empowerment of people by favouring integrity (a virtue ethic, which is about practising what one preaches and believes is right) and morality (the outward manifestation of a moral position), apart from their continuous professional development, would lead to a workforce that can achieve consistent superior performance on projects. Except for 'indifference', the HQT aspects illustrated in Figure 1.2 express the kind of treatment that is proper for satisfactory people-based human conditions.

In addition to the HQT aspects that are depicted in Figure 1.2, the notion of quality of working life (QWL) resonates with the well-being of workers and corresponding job satisfaction. The quality of the working conditions within a firm can influence the quality of work, productivity, and morale in a workplace. QWL has been described as the quality of the relationship between workers and their working environment where human dimensions are aligned with technical and economic considerations (Gallie, 2009; Nadler & Lawler, 1983; Sirgy, Efraty, Siegel & Lee, 2001). The reasoning behind QWL is that people will be more productive when they actively enjoy their work experience (Naoum, 2011). The principles of QWL have been captured by HQT and are included in Figure 1.2. Among others, QWL seeks to change the culture of a workplace by promoting stress reduction programmes, improving working conditions and advancing the use of life-long learning tools to motivate workers.

What does being people in construction mean?

Understanding the idea of 'what being human in construction means' requires unpacking issues around human ontological projects. Several authors have uncovered immoral and unethical acts and conditions in the construction industry. Ethical issues that have made headlines include corruption, bribery, fronting, and collusion, apart from dismal H&S performance of the industry. These acts, which are driven by profits, have significant impact on individual workers and their employers. The impact is enough for workers and construction firms to consider the wider harm from their work and product on themselves and society. The idea underpins such considerations as what is good for a person must also be good for humanity (Fromm, 1947). This idea from Fromm has been quoted and illustrated in construction empirical publications on ethical reasoning (McAleenan & McAleenan, 2011; 2012; 2013; 2014). Box 1.4 contends that respect for the dignity of others is the first step in understanding what it means to be human in construction. Starting with the need to clearly define what human is, Box 1.4 goes on to emphasise the need to look beyond Kant's deontology to how production work is enacted and what is viewed as fair remuneration in the construction industry. For example, production processes and work methods in the construction sector should move away from a myopic focus on responsive accident-prevention activities (McAleenan & McAleenan, 2014). Rather, workplaces should promote the creation of a culture that enhances the wellbeing and welfare of workers. After all, the ILO has always reiterated the notion that health, safety, and well-being are fundamental human rights that are enshrined in the United Nations Declaration on Human Rights of 1948. However, for the industry to appreciate and apply the tenets of fundamental human rights, it must understand what human nature is and how people determine what is good (and not good) for individuals and themselves (McAleenan & McAleenan, 2014). A comprehension of what is right and what is evil will lead to the development of appropriate ethical reasoning that will influence the approach to work in the construction industry. More so, Kantians hold the view that the actual worth of rational nature (humanity) is presupposed by the ability to act for the sake of expected values. The expectations of people in construction should mirror their actual worth, and everyone working in the industry should be expected to act in ways that do not violate the values of individuals.

Summary

This chapter has illustrated what characterises people in the operational phase of construction. Everyone who has to make contributions on site to realise a construction product is affected by the discourse in the chapter. These sets of people have their 'boots on the ground', while functioning as managers, professionals, technicians, craftspersons and workers in the industry. Their engagement is based on atypical and typical employment practices that often define the nature of their contributions in the construction process. The different employment regimes do not, however, isolate people from influences of the nature of construction work. This realisation suggests that people are exposed to many types of hazards which could potentially harm them if the required GPP is not applied. Thus, rather than merely prioritising the health, safety and well-being of people, it is time for the industry to value them based on morality (and not necessarily based on the business case).

The chapter has therefore argued that people must be treated with the human dignity, respect, authentic care and justice that foster their development. An indifferent treatment of individuals is not acceptable, and such treatment has an

Box 1.4 Being human in construction

To determine why we must respect the dignity of others and to accord equal consideration to all, it is necessary that we determine what it is to be human rather than decide who the people in construction are. To value people, we must value them as humans firstly, and then as construction workers.

To get to the root of humanity, it is necessary to distinguish that which makes humans from that which makes non-humans. There are two features, namely species awareness, i.e. we are aware of ourselves as a species as distinct from other species, and from this empathy develops, in that we see ourselves in others and others in ourselves. This is a foundation of moral behaviour, i.e. our relations with and behaviour towards others.

The second condition is that we are historical beings, i.e. we not only exist in time, but we are aware that we exist in time. This means that we live in the 'now' and are aware of the past and can anticipate the future. In this respect, we have the capacity to critically assess the past and how we have arrived at now, and from that, we can, in anticipation of a particular future, decide on and act towards that end.

In the complex social reality people struggle to attain their authentic humanity in the face of power relations, social constructs and ideological negations of our critical thinking and decision-making capacity. This is the human ontological project, the struggle to assert our Being, our authentic Self within the milieu of social relationships and the cultural environment we inhabit.

Kant's deontology, though framed as a means of being human, is a negation of that essential element of humanness, namely critical thought and agency. The categorical imperative limits adaptability in different circumstances: it creates inflexibility and directs individual behaviour down a particular route regardless of whether the circumstances require it. Deontology as rules and regulations, particularly absolute rules, negates thinking and thus agency. These rules thus deny our own humanity by abstracting the individual from his/her true being, no longer permitting critical reflection and autonomous decision making about what to do next.

Kant's deontology, while permitting of fair remuneration as an acceptable means of respecting workers, was none the less developed before modern theories of political economy. Wage labour, within which pay, fair or otherwise, exists, is an alienating mode of production, wherein the fruit of a people's labour does not belong to them, nor, as a consequence of the work contract, do they have decision-making capacity about what they produce, how they produce it and where the product of their labour ends up. All is outside their control, and no matter how good the labour conditions may be made in construction, it remains alienating work by the nature of the mode of production.

Dialogue authored by Philip McAleenan

adverse impact on the well-being of the workforce and the image of the industry. Inhuman treatment of people is unethical and counterproductive. The underlying idea behind Kantian ethics of morality provides the basis for respecting every worker in construction properly without breaches that would compromise the livelihood of people. Instead of the mechanistic view of the organisation that is more or less a tradition in the construction industry, it is argued that people are more than interchangeable parts or resources in an enterprise. In other words, the industry should not use people to get what it wants from them (use people as ends, not as means to an end). The practice of construction management should go beyond methods, processes, and technology by considering the 'humanity' in the workforce in order to improve the image and output of the industry.

The chapter has also signposted key issues in the book. Using sense-making ideas, the reader should reflect upon the critical perspectives provided in subsequent chapters. The nuances that impact on how people are treated and valued (or not valued) in the industry should encourage readers to join the growing calls to action regarding attracting and retaining people in the construction sector.

References

Abdelhamid, T. S. & Everett, J. G. (2000), Identifying root causes of construction accidents, *Journal of Construction Engineering and Management*, Vol. 126(1), pp. 52–60.

Adsul, B. B., Laad, P. S., Howal, P. V. & Chaturvedi, R. M. (2011), Health problems among migrant construction workers: A unique public–private partnership project, *Indian Journal of Occupational and Environmental Medicine*, Vol. 15(1), pp. 29–32.

Aires, M. D. M., Gámez, M. C. R. & Gibb, A. (2010), Prevention through design: The effect of European directives on construction workplace accidents, *Safety Science*, Vol. 48(2), pp. 248–258.

Andorno, R. (2009), Human dignity and human rights as a common ground for a global bioethics, *Journal of Medicine and Philosophy*, Vol. 34(3), pp. 223–240.

Arbulu, R., Tommelein, I., Walsh, K. & Hershauer, J. (2003), Value stream analysis of a re-engineered construction supply chain, *Building Research & Information*, Vol. 31(2), pp. 161–171.

Beavers, J., Moore, J., Rinehart, R. & Schriver, W. (2006), Crane-related fatalities in the construction industry, *Journal of Construction Engineering and Management*, Vol. 132(9), pp. 901–910.

Bust, P. D., Gibb, A. G. & Pink, S. (2008), Managing construction health and safety: Migrant workers and communicating safety messages, *Safety Science*, Vol. 46(4), pp. 585–602.

Carter, G. & Smith, S. D. (2006), Safety hazard identification on construction projects, *Journal of Construction Engineering and Management*, Vol. 132(2), pp. 197–205.

Chi, C.-F., Yang, C.-C. & Chen, Z.-L. (2009), In-depth accident analysis of electrical fatalities in the construction industry, *International Journal of Industrial Ergonomics*, Vol. 39(4), pp. 635–644.

Chi, S., Han, S. & Kim, D. Y. (2012), Relationship between unsafe working conditions and workers' behavior and impact of working conditions on injury severity in US construction industry, *Journal of Construction Engineering and Management*, Vol. 139(7), pp. 826–838.

Cranor, C. F. (1983), On respecting human beings as persons, *The Journal of Value Inquiry*, Vol.17(2), pp. 103–117.

Dainty, A. & Loosemore, M. (2013), *Human resource management in construction: Critical Perspectives*, Routledge, Abingdon.

Dainty, A., Green, S. & Bagilhole, B. (2007), People and culture in construction, *People and Culture in Construction: A reader*: 1, Routledge, Abingdon.

Darwall, S. L. (1977), Two kinds of respect, *Ethics*, Vol. 88(1), pp. 36–49.

De George, R. T. (2011), *Business ethics*, Pearson Education, India.

Dekker, S. W. (2014), The bureaucratization of safety, *Safety Science*, Vol. 70, pp. 348–357.

Dias, L. A. (2009), *Inspecting occupational safety and health in the construction industry*, International Training Centre of the International Labour Organization, Turin.

Dong, X. & Platner, J. W. (2004), Occupational fatalities of Hispanic construction workers from 1992 to 2000, *American Journal of Industrial Medicine*, Vol. 45(1), pp. 45–54.

Donnelly, J. (2013), *Universal human rights in theory and practice*, Cornell University Press, New York.

Edwards, D. & Nicholas, J. (2002), The state of health and safety in the UK construction industry with a focus on plant operators, *Structural Survey*, Vol. 20(2), pp. 78–87.

Egan, J. (1998), Rethinking construction. *Construction Task Force Report for Department of the Environment, Transport, and the Regions*, HMSO, London.

Forbes, L. H. & Ahmed, S. M. (2010), *Modern construction: Lean project delivery and integrated practices*, CRC Press, Boca Raton.

Fromm, E. (1947), *Man for himself*, Routledge, Oxford.

Gallie, D. (2009), *Employment regimes and the quality of work*, Oxford University Press, Oxford.

Gervais, M. (2003), Good management practice as a means of preventing back disorders in the construction sector, *Safety Science*, Vol. 41(1), pp. 77–88.

Ghaemi, H. (2006), Building towers, cheating workers: Exploitation of migrant construction workers in the United Arab Emirates, *Human Rights Watch* November 2006, p. 18.

Glendon, A. I. & Litherland, D. K. (2001), Safety climate factors, group differences and safety behaviour in road construction, *Safety Science*, Vol. 39(3), pp. 157–188.

Glendon, M. A. (1997), Knowing the Universal Declaration of Human Rights, *Notre Dame Law Reiew*, Vol. 73, p. 1153.

Halpin, D. W. & Riggs, L. S. (1992), *Planning and analysis of construction operations*, John Wiley & Sons, New York.

Haslam, R. A., Hide, S. A., Gibb, A. G., Gyi, D. E., Pavitt, T., Atkinson, S. & Duff, A. (2005), Contributing factors in construction accidents, *Applied Ergonomics*, Vol. 36(4), pp. 401–415.

Holt, A. S. J. (2008), *Principles of construction safety*, John Wiley & Sons, New York.

Janicak, C. A. (2008), Occupational fatalities due to electrocutions in the construction industry, *Journal of Safety Research*, Vol. 39(6), pp. 617–621.

Jeong, B. Y. (1998), Occupational deaths and injuries in the construction industry, *Applied Ergonomics*, Vol. 29(5), pp. 355–360.

Johnson, O. A. (1974), The Kantian interpretation, *Ethics*, Vol. 85(1), pp. 58–66.

Kant, I. (2012), *Fundamental principles of the metaphysics of morals*, Courier Corporation, North Chelmsford, MA.

Kant, I. & Friedman, M. (2004), *Kant: Metaphysical foundations of natural science*, Cambridge University Press, Cambridge.

Kisner, S. M. & Fosbroke, D. E. (1994), Injury hazards in the construction industry, *Journal of Occupational and Environmental Medicine*, Vol. 36(2), pp. 137–143.

Koskela, L. (1997), Lean production in construction, *Lean Construction*, pp.1–9.

Kuruvila, M., Dubey, S. & Gahalaut, P. (2006), Pattern of skin diseases among migrant construction workers in Mangalore, *Indian Journal of Dermatology, Venereology, and Leprology*, Vol. 72(2), pp. 129–132.

Langford, D. & Agapiou, A. (2007) The impact of eastward enlargement on construction labour markets in European Union member states. In Dainty, A., Green, S. & Bagilhole, B. (eds), *People and culture in construction: A reader*. pp. 205–221. Routledge, Abingdon.

Latham, S. M. (1994), *Constructing the team*, HM Stationery Office, London.

Lee, P. & George, R. P. (2008), The nature and basis of human dignity, *Ratio Juris*, Vol. 21(2), pp. 173–193.

Ling, F. Y. Y., Liu, M. & Woo, Y. C. (2009), Construction fatalities in Singapore, *International Journal of Project Management*, Vol. 27(7), pp.717–726.

Lingard, H., Brown, K., Bradley, L., Bailey, C. & Townsend, K. (2007), Improving employees' work-life balance in the construction industry: Project alliance case study, *Journal of Construction Engineering and Management*, Vol. 133(10), pp. 807–815.

Lingard, H. & Francis, V. (2004), The work-life experiences of office and site-based employees in the Australian construction industry, *Construction Management and Economics*, Vol.22(9), pp. 991–1002.

Lipscomb, H. J., Glazner, J. E., Bondy, J., Guarini, K. & Lezotte, D. (2006), Injuries from slips and trips in construction, *Applied Ergonomics*, Vol. 37(3), pp. 267–274.

Loushine, T. W., Hoonakker, P. L., Carayon, P. & Smith, M. J. (2006), Quality and safety management in construction, *Total Quality Management and Business Excellence*, Vol. 17(9), pp. 1171–1212.

McAleenan, P. & McAleenan, C. (2011), Enhancing ethical reasoning in design education. *Proceedings International Council for Building (CIB) W099 Conference*, August 2011, Washington, DC, USA.

McAleenan, C. & McAleenan, P. (2012), The degree of sophistication of ethics reasoning amongst first year under-graduate students. *Proceedings International Council for Building (CIB) W099 Conference*, September 2012, Singapore.

McAleenan, P. & McAleenan, C. (2013), Maturing workplace culture in the context of evolved ethical agency. *Proceedings International Council for Building (CIB) World Building Congress*, May 2013, Brisbane, Australia.

McAleenan, C. & McAleenan, P. (2014), The application of deductive logic to determine the objective conditions impacting upon cultural maturity. *Proceedings International Council for Building (CIB) W099 Conference*, May 2014, Lund, Sweden.

McGeorge, D. & Zou, P. (2013), *Construction management: New directions* (3rd edn). Wiley-Blackwell, Oxford.

Melé, D. (2014), "Human quality treatment": Five organizational levels, *Journal of Business Ethics*, Vol. 120(4), pp. 457–471.

Murray, M., Chan, P. & Tookey, J. (2001), Respect for people: Looking at KPI's through 'younger eyes'! In: A. Akintoye, (ed.). Proceedings 17th Annual Association of Researchers in Construction Management (ARCOM) Conference, 5–7 September 2001, Salford, UK. *Association of Researchers in Construction Management*, Vol. 1, pp. 671–80.

Nadler, D. A. & Lawler, E. E. (1983), Quality of work life: Perspectives and directions, *Organizational Dynamics*, Vol.11(3), pp. 20–30.

Naoum, S. (2011), *People and organizational management in construction* (2nd edn), Thomas Telford, London.

Ness, K. (2010), The discourse of 'Respect for People' in UK construction, *Construction Management and Economics*, Vol. 28(5), pp. 481–493.

Ness, K. & Green, S. (2008), Respect for people in the enterprise culture? *Paper presented at the 24th Annual Association of Researchers in Construction Management (ARCOM) Conference*, September 2008, Cardiff.

Paap, K. (2006), *Working construction*, ILR Press, Ithaca.

Pinto, A., Nunes, I. L. & Ribeiro, R. A. (2011), Occupational risk assessment in construction industry – Overview and reflection, *Safety Science*, Vol. 49(5), pp. 616–624.

Respect for People Working Group. (2004), *Respect for people: A framework for action*, Constructing Excellence, London.

Rethinking Construction, R. (2000), A commitment to people: "Our biggest asset". *A report from the Movement for Innovation's working group on Respect for People*, Rethinking Construction, London.

Ringen, K., Seegal, J. & England, A. (1995), Safety and health in the construction industry, *Annual Review of Public Health*, Vol. 16(1), pp. 165–188.

Sang, K., Dainty, A. & Ison, S. (2007). Warning: Working in construction may be harmful to your psychological well-being, *People and culture in construction: A reader*, Taylor & Francis, New York.

Senge, P. M. (2006), *The fifth discipline: The art and practice of the learning organization*, Broadway Business, Random House, London.

Sirgy, M. J., Efraty, D., Siegel, P. & Lee, D.-J. (2001), A new measure of quality of work life (QWL) based on need satisfaction and spillover theories, *Social Indicators Research*, Vol.55(3), pp. 241–302.

Taylor, P. W. (2003), The ethics of respect for nature, *Environmentalism: Critical Concepts*, Vol. 1(3), p. 61.

Teo, M. & Loosemore, M. (2001), A theory of waste behaviour in the construction industry, *Construction Management & Economics*, Vol. 19(7), pp. 741–751.

Toole, T. M. (2002), Construction site safety roles, *Journal of Construction Engineering and Management*, Vol. 128(3), pp. 203–210.

Trajkovski, S. & Loosemore, M. (2006), Safety implications of low-English proficiency among migrant construction site operatives, *International Journal of Project Management*, Vol. 24(5), pp. 446–452.

Watts, J. H. (2009), 'Allowed into a man's world': Meanings of work–life balance: Perspectives of women civil engineers as 'minority' workers in construction, *Gender, Work & Organization*, Vol. 16(1), pp. 37–57.

2 Construction workers' health

Occupational and environmental issues

Helen Lingard and Michelle Turner

Have the courage to use your own reason – that is the motto of enlightenment
(Immanuel Kant)

Background

Traditionally, occupational health and safety (OHS) research has focused heavily on the prevention of accidents that give rise to acute effect injuries. However, far less attention has been paid to the issue of construction workers' health.

Good health contributes to economic and social well-being, while poor health keeps people from working and causes significant hardship (Welch, 2009). Longer life expectancy and declining birth rates are combining to create an ageing population in nearly all countries in the world. The United Nations estimates that the global share of older people (aged 60 years or over) grew from 9.2 per cent in 1990 to 11.7 per cent in 2013 and will reach 21.1 per cent by 2050. The old-age support ratio, i.e. the number of working-age adults per older person in the population, is low in developed and developing countries and is projected to fall further with significant pressure on social and economic support systems (United Nations, 2013). Labour participation rates decline significantly after the age of 55 years. For example, in Australia, there is a 10 per cent reduction in labour force participation rates for males and females over the transition from 55 to 59 to 60 to 64, and a further 20 per cent decrease over the next five-year age range (Australian Government Productivity Commission, 2013). Governments are actively trying to increase workforce participation among older workers. However, research shows that many construction workers retire early owing to poor health or work disability (Brenner & Ahern, 2000; Welch, 2009; Oude Hengel et al., 2012).

Construction workers are a disadvantaged socio-economic group regarding job conditions, levels of education and wealth/assets. Research shows that adult men whose behaviour and psychosocial disposition have negative health impacts are more likely to:

- have experienced poor childhood conditions;
- have low levels of education; and
- work in blue-collar employment (Lynch et al., 1997).

Kolmet et al. (2006) used a socio-ethnographic approach to understand the determinants of health among Australian male blue-collar workers and found that health expectations are low owing to anticipated 'wear and tear' caused by the physical demands of their work, stress inherent in balancing work and family demands, and overall lifestyle. Even compared to other blue collar occupations, construction workers experience high levels of work incapacity (Arndt et al., 2005) and, in some countries, up to two-thirds of construction workers retire early owing to permanent disability (Siebert et al., 2001).

Perspectives and concepts

Occupational health risks

Construction work is physically demanding, and workers are exposed to many occupational health hazards, including harmful substances (e.g. clouds of dust, fumes, gases, toxic chemicals), biological hazards (e.g. acute and chronic infections, parasites) and physical hazards (e.g. noise, heat, cold, vibration) (Snashall, 2005).

Construction workers experience elevated rates of contact dermatitis, all types of skin neoplasm, non-malignant pleural disease, mesothelioma, lung cancer, pneumoconiosis and musculoskeletal disorders (Stocks et al., 2010; Stocks et al., 2011). In the UK, over half of the occupational attributable cancer deaths are the result of exposures within construction (Rushton et al., 2008). Cancers caused by exposure to asbestos in building products have been widely documented (LaDou, 2004). The World Health Organization (WHO) estimates that currently about 125 million workers are exposed to asbestos, and more than 107,000 will die each year from asbestos-related lung cancer, mesothelioma, and asbestosis (WHO, 2016). While the use of asbestos in industrial products has been banned or seriously restricted in many developed countries, asbestos is still used in many parts of the world, often with little attempt made to protect workers from harm (Kazan-Allen, 2005). It is therefore highly probable that a new 'wave' of asbestos-related deaths will occur in developing countries in the future.

In addition to the demonstrated links between asbestos exposure and cancer, some commonly used construction products have also been linked to a high prevalence of cancer. For example:

(i)   cement dust, which has been linked to throat cancer (Purdue et al. 2006; Dietz et al. 2004) and oesophageal cancer (Jansson et al., 2005);
(ii) asphalt fumes, which have been linked to cancers of the gastric system (Jansson et al., 2005); and
(iii) wood dust, linked to cancers of the nose (Siew et al., 2012).

Furthermore, construction workers' exposure to ultraviolet light and chemicals also makes them a high-risk group for skin cancer (Pritchard & Dixon, 2008; Håkansson et al., 2001).

Psychological risks and mental health

Project-based construction work is characterised by long hours of work, job insecurity, perceived conflict between work and family life and low levels of autonomy and control. All of these work characteristics have been linked to diminished health (Hannertz et al., 2005; Van der Hulst, 2003; Turner & Lingard, 2016). Research shows that construction workers are a high-risk group for psychological distress and mental illness (Abbe et al., 2011). Job burnout has been reported to be high in managerial and professional construction workers (Lingard & Francis, 2005; Lingard et al., 2007). High levels of burnout leading to early retirement have also been reported in manual, non-managerial workers (Oude Hengel et al., 2012). The incidence of mental distress among construction workers is reported to be twice the level of the general male population (Borsting Jacobsen et al., 2013). Most disturbingly, Meltzer et al. (2008) report a high rate of suicide among construction workers compared to other occupations. In Australia, construction workers are six times more likely to die by suicide than through a workplace accident, and construction apprentices are two and a half times more likely to commit suicide than other young men their age (http://www.matesinconstruction.org.au, 2016).

Links between stressors and physical injury

Construction workers are a high-risk group for musculoskeletal disorders (Petersen & Zwerling, 1998). In Australia, back injuries account for 20 per cent of serious compensation claims in construction workers, while knee and upper leg injuries and injuries to the shoulder and upper arm account for 12 and 10 per cent respectively (Safe Work Australia, 2016). Manual, non-managerial construction work involves frequent exposure to awkward posture and bodily movements including lifting, bending, twisting or kneeling, often for extended periods (Valero et al., 2016). Latza et al. (2002) report that the repetitive laying off massive sandstones for more than two hours per shift significantly increases the risk of chronic lower back pain with a risk advancement period of 18 years. This means that a 40-year-old performing this task has the same risk as a 58-year-old construction worker not performing this task. The long-term impacts of musculoskeletal injuries are significant. In the USA, 40 per cent of construction workers over the age of 50 are reported to experience chronic back pain (Dong et al., 2012).

Traditional approaches to health have viewed the mind and body as functioning independently. This view does not adequately explain illness, which refers to a person's subjective appraisal of the symptoms of disease (Gatchel et al., 2007). There is a growing understanding that psychological and social processes interact with biological factors in shaping mental and physical health. Consistent with this view, there is a substantial body of research evidence demonstrating a link between the experience of workplace psychosocial stressors (for example, low levels of social support and job satisfaction and high perceived job stress) and musculoskeletal problems (Hoogendoorn et al., 2000; Bongers et al., 2002).

Borsting Jacobsen et al. (2013) report that mental distress in construction workers is strongly significantly associated with the experience of lower back pain, having two or more pain sites and the experience of injury. Stattin and Järvholm (2005) similarly report that physical, ergonomic and psychosocial work demands all increase the odds ratio of construction workers seeking a disability pension. Importantly, the effects of physical and ergonomic risk factors are exacerbated when workers have low job control or autonomy. Similarly, a Danish study reveals psychosocial work factors predict the early retirement of workers in physically heavy occupations, even after implementing control measures for disease (Lund et al., 2001).

The pathway between psychosocial risk factors and chronic pain is not entirely clear, but some possible explanations for the links have been suggested as follows.

- Demanding work directly impacts the speed and acceleration of movement, applied force and posture, contributing to musculoskeletal injury.
- Demanding work triggers sustained stress responses and may cause physiological changes contributing to musculoskeletal injury.
- Stress responses may lead workers to appraise work situations and musculoskeletal symptoms differently.
- Stress responses may influence the transition from acute to chronic musculoskeletal pain (Bongers et al., 2002).

This research points to the potential benefits of providing an integrated preventive approach focused on addressing both physical and psychosocial risk factors inherent in construction work (MacDonald et al., 2001).

The occupational health imperative

The regulation of occupational health and safety establishes responsibilities for employers, employees, and other duty holders. In the United Kingdom and many other countries that have adopted the principle-based regulatory model, the legislation requires duty holders to reduce risks to workers' health and safety to be as low as reasonably practicable (ALARP). Underpinning the ALARP principle is a hierarchy of control (HOC) that classifies ways of controlling hazards according to their level of effectiveness. At the top of the HOC is the elimination of a hazard altogether. This is the most effective form of control because a hazard is physically removed from the workplace and no longer poses a threat to health or safety. If elimination is not possible, the next best option for risk control is a substitution. This involves replacing something that produces a hazard with something less hazardous. Beneath substitution in order of preference are engineering controls, i.e. controls that physically isolate workers from hazards. The top three layers of the HOC (i.e. elimination, substitution, and engineering) are classed as technological in nature because they change the physical work environment. In contrast, the bottom two layers of the HOC represent behavioural controls that seek to change the way people work. These are administrative controls, which include such measures as implementing safe work procedures or a job rotation scheme to limit exposure to a hazard. The lowest level of control in the HOC is personal protective equipment (PPE).

While HOC is usefully applied to address dangers of acute-effect safety issues, Sherratt (2015) observes that it is less effectively applied to the elimination, containment or control of health risks. The extent to which the construction industry is effectively managing occupational health hazards under this framework has been questioned (Gyi et al., 1998).

Public versus occupational health

Driven by a public health agenda, health promotion programmes seek to improve wellness beyond the workplace and often focus on issues of workers' health-related behaviours. Policy statements articulating the benefits to be gained from improving workers' health and well-being focus heavily on the 'business case'. Appeals are made to managers to invest in workers' health and well-being to improve productivity and reduce absenteeism, compensation claims, and voluntary turnover. For example, WorkSafe Victoria's Healthy Workers' Kit cites the following evidence.

- Every US$1 invested in health and well-being can achieve a return of US$5.82 in reduced absenteeism costs.
- Workplace health programmes can reduce sick leave by up to 30 per cent and increase productivity by up to 52 per cent.
- Workplace health programmes can achieve an average of a 32 per cent reduction in workers' compensation and disability claim costs (WorkSafe Victoria, 2016).

Such 'calls to action' position corporate interest as the principal motive for protecting and promoting workers' health. The focus on workers' health behaviours outside of work can also obfuscate the effects of exposure to occupational health risks, which organisations are legally bound to reduce to ALARP.

Behaviour-change programmes

So-called 'lifestyle' diseases (obesity, cardiovascular disease, and the like) have been identified as factors contributing to work disability (Alavinia et al., 2007, Claessen et al., 2009). Consequently, publicly-funded health programmes have been implemented to improve or change the health-related behaviour of construction workers (Ludewig & Borstad, 2003; Groeneveld et al., 2010). In some cases, these programmes have produced some positive improvements in changing diet, eating habits and physical exercise behaviours (Sorensen et al., 2007; Gram et al., 2012).

Sherratt (2015) argues that such programmes are paternalistic attempts to place constraints on workers' freedom and choice. Philosophical questions about the extent and (il)legitimacy of corporate interference into workers' private lives deserve serious consideration. However, a more basic concern lies in the effectiveness of behavioural health promotion programmes in construction.

Methodology

Lingard and Turner (2015) report on a participatory action research project conducted at three construction projects in Australia. The research was funded by the Department of Justice and Attorney General under the Queensland Government *Healthier. Happier. Workplaces* initiative. The initiative specifically focused on five health risk factors (i.e. smoking, poor nutrition, excessive alcohol intake, physical inactivity, and obesity). Each worksite independently developed health promotion activities during the research. Activities implemented included:

(1) the introduction of a smoking cessation programme that was available to workers and their family members;
(2) a healthy eating promotion programme that included food tasting, cooking instruction and the provision of healthy food options in the site canteen;
(3) the introduction of on-site yoga and stretching sessions;
(4) the introduction of a competitive programme to encourage workers to become more physically active.

In addition to this, fresh fruit was delivered to the site on a weekly basis.

A baseline survey was undertaken with workers at participating construction sites at the beginning of the research period. The Australian version of the SF-36 instrument was used to measure health status. The SF-36 is a generic, multi-purpose short form survey that produces an eight-scale profile of health and well-being (Ware, 1999). The eight domains of the SF-36 are:

(1) Physical Functioning (PF): the presence and extent of physical limitations;
(2) Role – Physical (RP): physical health-related role limitations;
(3) Bodily Pain (BP): level of pain and impact of normal work duties;
(4) General Health (GH): evaluation of general health;
(5) Vitality (VT): energy level and fatigue;
(6) Social Functioning (SF): impact of either physical or emotional problems on social activities;
(7) Role – Emotional (RE): impact on work as a result of emotional problems; and
(8) Mental Health (MH): evaluation of mental health.

The SF-36 is widely used and accepted as a reliable measure of health status (Garratt et al., 2002; McDowell & Newell, 1996). Furthermore, the construct, criterion and content validity of the SF-36 has been demonstrated (Maruish, 2011). Scores for all health domains are expressed on a scale of 0 to100, where a higher score indicates a better state of health or well-being. For five of the domains (physical functioning, role-physical, bodily pain, social functioning, and role-emotional), the highest possible score of 100 indicates the absence of health limitations or disabilities. For the other three domains (general health, vitality, and mental health), a score of 100 indicates a positive state of well-being.

Results

Table 2.1 shows that scores for vitality and social functioning (two of the domains that feed into the mental health component of the SF-36) were lower than the Australian male norm scores for all age groups except the 60 years and over a group. Construction workers aged between 30 to 39 years of age reported lower scores that the Australian male norm scores for all of the SF-36 component scales except for physical functioning. Scores for all components were lower among construction workers aged between 30 and 39 than they were among construction workers in younger and older adjacent age brackets.

Workers at participating construction sites were also asked to complete weekly logs which recorded health behaviour relating to the consumption of healthy and unhealthy foods, and physical activity. Figures 2.1, 2.2 and 2.3 show data collected over 18 weeks at the largest of the participating worksites. Figure 2.1 shows fluctuations in daily serves of fruit consumption; Figure 2.2 shows the number of days per week junk food was eaten, and Figure 2.3 shows the level of physical activity of 30 or more minutes per week outside of work. Moving average analysis was applied to the data to explore whether there were changes in health behaviour during the 18-week period. No significant changes were found, suggesting that the interventions implemented at the site did not change workers' health behaviour (Turner & Lingard, 2016).

Table 2.1 Health domain scores by age (construction sample)

Health domain	Under 30 (n = 35)	30-39 (n = 50)	40-49 (n = 28)	50-59 (n = 25)	60 and over (n = 4)
Physical functioning	96.4	93.3	95.5	83.8	80.0
Role – physical	88.2	84.1	93.5	86.2	90.6
Bodily pain	79.9	72.6	78.7	70.8	79.5
General health	72.8	61.3	71.3	63.1	77.2
Vitality	62.6	54.1	63.8	51.8	59.3
Social functioning	82.5	71.7	83.4	80.0	100.0
Role – emotional	90.0	80.2	91.0	87.3	95.8
Mental health	78.4	68.2	76.5	72.3	82.5

NB: Figures in bold indicate that the domain score is lower than the equivalent Australian male age-based population norm.

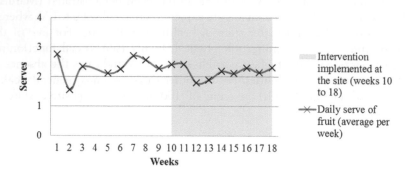

Figure 2.1 Serves of fruit per day

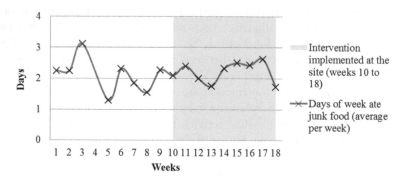

Figure 2.2 Number of days per week junk food eaten

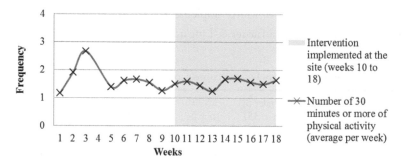

Figure 2.3 Level of physical activity of 30 or more minutes per week outside of work

In order to explore reasons why no sustained or significant improvements in health-related behaviour were observed, workers at participating work-sites were interviewed and/or participated in focus group sessions. These sessions provided rich explanatory data that revealed systemic causes of unhealthy behaviour, driven by long hours of work, job stress, concerns about job and financial security and work-family conflicts. Long work hours and structural characteristics of employment in the construction industry are also reported by Townsend et al. (2016) to be an impediment to healthy eating.

Implications for practice and research

Complex causes of poor health and work disability

The Queensland research described above highlights a fundamental challenge associated with addressing workers' behaviour without paying due consideration to work organisation and environmental conditions. Lingard and Turner (2015) used an ecological perspective to explain their findings. Ecological health theorists place increased focus on people's contexts, seeking to identify environmental constraints on behaviour and thus progress beyond a linear, mechanistic way of understanding the causes of poor health (McLaren & Hawe, 2005). Proponents of an ecological perspective suggest that some individual or environmental conditions have a disproportionately high level of influence on individuals' health and well-being (Grzywacz & Fuqua, 2000). Stokols et al. (1996) observe that health promotion approaches that focus solely on changing behaviours often neglect the contextual factors that produce high rates of relapse and attrition once specific periods of intervention come to an end. In contrast, ecological approaches to workplace health promotion are designed to address the complex interdependencies between individual determinants of health and aspects of the physical, social, cultural and organisational work environments (Stokols et al. 2003). An ecological perspective suggests that targeting workers' behaviour without also addressing environmental determinants of health is likely to produce weak or short-lived results (Sallis et al., 2008).

Consistent with this view, Noblet and LaMontagne (2006) call for more comprehensive approaches to the design of workplace health promotion programmes, which address adverse conditions of work at the same time as workers' health behaviours. This is consistent with policies designed to integrate occupational health and safety programmes with workplace health promotion programmes, such as the US National Institute of Occupational Safety and Health (NIOSH) Total Worker Health programme (Schill & Chosewood, 2013).

Health and the quality of work

The extant research highlights the need to address the quality of work in the design and implementation of health promotion programmes. The quality of work is known to impact workers' health (Butterworth et al., 2011). In particular, some of the environmental, psychosocial risk factors to which construction workers are exposed are likely to have direct health impacts, and indirectly affect health through shaping poor health behaviours. For example, long work hours and poor work–family balance are linked to:

- poor diet (Devine et al., 2006; Devine et al., 2007);
- lack of physical exercise (Burton & Turrell, 2000; Van Steenbergen & Ellemers, 2009); and
- harmful levels of alcohol consumption (Frone et al.,1997; Roos et al., 2006).

The *Australian Strategy for Work Health and Safety* 2012–2022 identifies health and safety by design as a priority action area. It specifies that hazards should be eliminated or minimised by design and, in particular:

- structures, plant, and substances should be designed to eliminate or minimise hazards and risks before they are introduced into the workplace; and
- work, work processes and systems of work should be designed and managed to eliminate or minimise hazards and risks.

In construction, there has been considerable focus on addressing workers' safety in the design of structures, but far less attention has been paid to the design of work itself. This is an important omission because of the demonstrated links between quality work and health. As the Australian Strategy states:

> Workers' general health and wellbeing are strongly influenced by their health and safety at work. Well-designed work can improve worker health.
> (Safe Work Australia, 2016)

The case study below describes an initiative of one large construction client organisation in New Zealand to create an integrated physical and psychosocial work environment that protects and promotes construction workers' health and well-being.

Case study: Client-driven health and well-being

The organisation implemented a programme to support *Safety-first, Quality-second, Time line-third* during an 18-month construction project. The agreement was made that health and safety would be a priority for all workers irrespective of whether they were directly employed by the organisation or engaged as contractors. A senior manager explained the organisation's approach to the project:

> We wanted to create a legacy that health and safety are the most important thing. We wanted our contractors to regain some type of life balance. There has been a history where contractors have worked hard, big hours. We wanted to give them the opportunity to regain life balance which would

have a positive impact on how they approached the project. We wanted also to strive towards creating a positive safety culture. We wanted to have a very clear picture of what was acceptable and what was not acceptable.

Leadership was recognised as critical to the success of the *Safety-first, Quality-second, Time line-third* initiative. To support this, the organisation implemented a leadership development training programme. A senior member of the project commented:

> We pulled in the key leaders from our main contractor groups and used that as a tool that starts to improve their performance, but also we took it as an opportunity within our construction management team to send some of our leaders to that.

Various skill sets were developed, such as conflict management. The development of leaders was considered successful:

> We have seen strengths and increases in performance particularly around climate surveys.

The organisation proactively managed the selection of leaders on the project, given this was identified as critical to the success of *Safety-first, Quality-second, Time line-third*. One of the senior managers explained:

> …we have changed some of the leadership team. So at certain points in the project we've recognised that certain leaders amongst the team aren't performing the way we want them to perform. And so we have asked for them to be either replaced or be supported by someone [to develop their capability].

A fatigue management policy (FMP) was implemented at the site, which stated that working time should not exceed 60 hours per week. Importantly, this policy was incorporated into the procurement strategy. At the front end of contract management, contractors were made aware of expectations of hours owing to the wellness focus at the project. Given the emphasis of the FMP, contractors were invited to negotiate time lines with the client to ensure that safety and quality standards were maintained. A senior manager explained:

> …rather than dictating to them a date that it [the project] had to be completed by, we negotiated with them as to how many people we can put on the work front and how that worked out to a timeline.

If more than 60 hours were planned, the organisation asked contractors to advise in advance how they would manage that week. The organisation challenged the contractors to consider extra resources to support the tail end of a week, and identify which activities could be brought to the front of the week so that only low-risk work was undertaken at the end of the week.

While a key aim of the policy was to prevent fatigue, it was also acknowledged that enforcing the policy enabled workers to spend time with their family. A worker commented:

> I've done 55 hours this week. I'm going home, and I'm going to spend the weekend with my family. I haven't done that before on a project.

Project culture was proactively developed and managed. An open and considered approach to communication, problem-solving and conflict management was applied. The organisation had purposively moved away from the *"policeman type role"* and taken on a consultative and educative approach. A senior manager commented:

> Lecturing people and hauling them over the coals over an incident isn't progressing the health and safety space for the project... instead of being proactive and working with them and finding better ways of doing things.

There was an emphasis on creating trust between the organisation and contractors. At the core of this was moving away from a blame culture and creating a culture which accepted open discussion and problem-solving. A senior manager on the project explained:

> ...when we did have incidents; we never were out to hang anyone, once again trying to keep that trust thing going.

To support the development of positive working relationships, the organisation established a village at the construction site. The aim of the village was to create an environment which *"...engages all the different contractor organizations to one central hub"*. Owing to space constraints, however, *"...there were areas of separation, but overall we were able to create a sense of a village environment where it was a welcoming place to come, and although we didn't have facilities like on other projects like cafeterias, we did have a social environment"*. A senior manager explained:

> ...we had barbecues and get-togethers, and it was an opportunity for the managers to talk to contractors about the values of the project and health and safety focus.

Throughout the duration of the project, communication was carefully considered. Messaging was kept simple and repetitive, and a variety of methods were used. For example, senior managers regularly went onsite and had safety conversations with workers. During high-stress times such as commissioning, strategies were implemented to create opportunities for communication such as daily meetings and barbecue lunches.

It was also acknowledged that the workers had limited time outside of work, so the organisation arranged for a nurse to attend the site on a regular basis.

Workers had free access to the nurse during work time, and this was considered important in supporting the health of workers. Furthermore, subject matter experts were engaged by the organisation to conduct targeted sessions in cases where an audit had identified a weakness. For example, an occupational health hygienist was brought onsite to present on hazardous chemicals, as an audit had identified a weakness. Attendance at this event was compulsory.

There were various benefits of the programme for the organisation, and some of these are summed up by a senior manager who reflected on what usually happens on a project, and how this project differed:

> Usually, the last two weeks before start up people are running in all directions and there's rubbish all over the floor, there are electrical cables, you know there's people stressed. At this project it looked like everything was calm and in control... people were having good quality conversations about how to install things, about how to wire something up, about how to weld. So you know that people are not fatigued. People have got time to think quietly about the best approach in how to install something. You know I saw a quality of workmanship, and I think that we've spent 75 million dollars on this project. I don't believe we've replaced one valve, pipe or instrument in 75 million dollars which is extraordinary. We were well-resourced as well, but a good part of that is that we look after people and when people are looked after they can think properly and think clearly and make good quality decisions.

Many contractors reported that it was the best project they had ever worked on owing to the focus on health and safety. A senior manager commented:

> ...we've had people coming up to us saying this is the best project that they've ever worked on and they've been in the industry for 15, 20 years.

The organisation aimed to leave a legacy that workers' health and safety is the most important factor on a project. The organisation challenged contractors to re-think their approach to construction-related activity, with a clear message that time line should not drive activity. A senior manager reflected:

> I think people will be talking about this project for a while. There's already people talking about changes to what they do now.

The above case study shows the positive effects that can be realised when attention is paid to the design of good quality work. Addressing some of the key psychosocial risk factors associated with construction work, such as long hours, low levels of social support, poor work-life balance and low levels of job autonomy or control can create a project-based environment which is supportive of workers' health.

Summary

Construction workers are a high-risk group for poor health, experiencing an elevated incidence of work-related disease and disability. In Australia, as in many other countries, employers have a legal responsibility to reduce workplace exposure to physical and psychosocial health risks to ALARP. These responsibilities must not be overlooked as the focus of health promotion programmes shifts to workers' health-related behaviours, both inside and outside the site environment. However, research has shown that interventions designed to change workers' health-related behaviour produce little or no improvement when they are introduced in an organisational or project environment that is not supportive of behaviour change. Thus, a more integrated approach to addressing workers' health is needed. Fundamental to improving the health of construction workers is the need to improve the quality of work, particularly for those in project-based roles. Given the highly competitive context of contracting and sub-contracting, it is highly likely that this is best achieved and driven by responsible client organisations.

Acknowledgement

Some of the research presented in this paper was funded by the Department of Justice and Attorney General under the Queensland Government 'Healthier. Happier. Workplaces' initiative and supported by Lendlease. We also gratefully acknowledge the assistance of Steve Nevin, Risk Manager, Assets and Capital Projects, Fonterra Co-operative Group Ltd.

References

Abbe, O.O., Harvey, C.M., Ikuma, L.H. & Aghazadeh, F. (2011), Modeling the relationship between occupational stressors, psychosocial/physical symptoms and injuries in the construction industry, *International Journal of Industrial Ergonomics*, Vol. 41(2), pp. 106–17.

AGPC *see* Australian Government Productivity Commission.

Alavinia, S. M., Van Duivenbooden, C. & Burdorf, A. (2007), Influence of work-related factors and individual characteristics on work ability among Dutch construction workers, *Scandinavian Journal of Work, Environment & Health*, Vol. 33(5), pp. 351–357.

Arndt, V., Rothenbacher, D., Daniel, U., Zschenderlein, B., Schuberth, S. & Brenner, H. (2005), Construction work and risk of occupational disability: A ten-year follow-up of 14,474 male workers, *Occupational and Environmental Medicine*, Vol. 62(8), pp. 559–566.

Australian Government Productivity Commission (AGPC). (2013), *An ageing Australia: Preparing for the future*. Commonwealth of Australia, Canberra.

Bongers, P. M., Kremer, A. M. & Laak, J. T. (2002), Are psychosocial factors, risk factors for symptoms and signs of the shoulder, elbow, or hand/wrist? A review of the epidemiological literature, *American Journal of Industrial Medicine*, Vol. 41(5), pp. 315–342.

Borsting Jacobsen, H., Caban-Martinez, A., Onyebeke, L., Sorensen, G., Dennerlein, J.T. & Endresen Reme, S. (2013), Construction workers struggle with a high prevalence of mental distress, and this is associated with their pain and injuries, *Journal of Occupational and Environmental Medicine*, Vol. 55(10), pp. 1197–204.

Brenner, H. & Ahern, W. (2000), Sickness absence and early retirement on health grounds in the construction industry in Ireland, *Occupational, and Environmental Medicine*, Vol. 57(9), pp. 615–620.

Burton, N. W. & Turrell, G. (2000), Occupation, hours worked, and leisure-time physical activity, *Preventive Medicine*, Vol. 31(6), pp. 673–681.

Butterworth, P., Leach, L. S., Strazdins, L., Olesen, S. C., Rodgers, B. & Broom, D. H. (2011), The psychosocial quality of work determines whether employment has benefits for mental health: Results from a longitudinal national household panel survey, *Occupational and Environmental Medicine*, Vol. 68(11), pp. 806–812.

Claessen, H., Arndt, V., Drath, C. & Brenner, H. (2009), Overweight, obesity and risk of work disability: A cohort study of construction workers in Germany, *Occupational and Environmental Medicine*, Vol. 66(6), pp. 402–409.

Devine, C. M., Jastran, M., Jabs, J., Wethington, E., Farrell, T. J. & Bisogni, C. A. (2006), 'A lot of sacrifices': Work–family spillover and the food choice coping strategies of low-wage employed parents, *Social Science & Medicine*, Vol. 63(10), pp. 2591–603.

Devine, C. M., Stoddard, A. M., Barbeau, E. M., Naishadham, D. & Sorensen, G. (2007), Work-to-family spillover and fruit and vegetable consumption among construction laborers, *American Journal of Health Promotion*, Vol. 21(3), pp. 175–81.

Dong, X. S., Wang, X., Fujimoto, A. & Dobbin, R. (2012), Chronic back pain among older construction workers in the United States: A longitudinal study, *International Journal of Occupational and Environmental Health*, Vol. 18(2), pp. 99–109.

Dietz, A., Ramroth, H., Urban, T., Ahrens, W. & Becher, H. (2004), Exposure to cement dust, related occupational groups and laryngeal cancer risk: Results of a population based case-control study, *International Journal of Cancer*, Vol. 108(6), pp. 907–911.

Frone, M. R., Russell, M. & Cooper, M. L. (1997), Relation of work–family conflict to health outcomes: A four-year longitudinal study of employed parents, *Journal of Occupational & Organizational Psychology*, Vol. 70(4), pp. 325–35.

Gatchel, R. J., Peng, Y. B., Peters, M. L., Fuchs, P. N. & Turk, D. C. (2007), The biopsychosocial approach to chronic pain: Scientific advances and future directions, *Psychological Bulletin*, Vol. 133(4), pp. 581–624.

Garratt, A., Schmidt, L., Mackintosh, A. & Fitzpatrick, R. (2002), Quality of life measurement: Bibliographic study of patient assessed health outcome measures, *British Medical Journal*, Vol. 324, pp. 1417–1419.

Gram, B., Holtermann, A., Sogaard, K. & Sjogaard, G. (2012), Effect of individualized worksite exercise training on aerobic capacity and muscle strength among construction workers – a randomized controlled intervention study, *Scandinavian Journal of Work, Environment & Health*, Vol. 38 (5), pp. 467–475.

Groeneveld, I. F., Proper, K. I., Van der Beek, A. J. & Van Mechelen, W. (2010), Sustained body weight reduction by an individual-based lifestyle intervention for workers in the construction industry at risk for cardiovascular disease: Results of a randomized controlled trial, *Preventive Medicine*, Vol. 51(3–4), pp. 240–246.

Grzywacz, J. G. & Fuqua, J. (2000), The social ecology of health: Leverage points and linkages, *Behavioral Medicine*, Vol. 26(3), pp. 101–115.

Gyi, D. E., Haslam, R. A. & Gibb. A. G. F. (1998), Case studies of occupational health management in the engineering construction industry, *Occupational Medicine* 48, 4, 263–271.

Håkansson, N., Floderus, B., Gustavsson, P., Feychting, M. & Hallin, N. (2001), Occupational sunlight exposure and cancer incidence among Swedish construction workers, *Epidemiology*, Vol. 12(5), pp.552–557.

Hannertz, H., Spangenberg, S., Tüchsen, F. & Albertsen, K. (2005), Disability retirement among former employees at the construction of the Great Belt Link, *Public Health*, Vol.119(4), pp. 301–304.

Hoogendoorn, W. E., Van Poppel, M. N., Bongers, P. M., Koes, B. W. & Bouter, L. M. (2000), Systematic review of psychosocial factors at work and private life as risk factors for back pain, *Spine*, Vol. 25(16), pp. 2114–2125.

Kazan-Allen, L. (2005), Asbestos and mesothelioma: Worldwide trends, *Lung Cancer*, Vol. 49, pp. S3–S8.

Kolmet, M., Marino, R. & Plummer, D. (2006), Anglo-Australian male blue collar workers discuss gender and health issues, *International Journal of Men's Health*, Vol. 5(1), pp. 81–91.

Jansson, C., Johansson, A. L., Bergdahl, I. A., Dickman, P. W., Plato, N., Adami, J., Boffetta, P. & Lagergren, J. (2005), Occupational exposures and risk of esophageal and gastric cardia cancers among male Swedish construction workers, *Cancer Causes & Control*, Vol. 16(6), pp. 755–764.

LaDou, J. (2004), The asbestos cancer epidemic, *Environmental Health Perspectives*, Vol. 112(3), pp. 285–290.

Latza, U., Pfahlberg, A. & Gefeller, O. (2002), Impact of repetitive manual materials handling and psychosocial work factors on the future prevalence of chronic low-back pain among construction workers, *Scandinavian Journal of Work, Environment & Health*, Vol. 28 (5), pp. 314–323.

Lund, T., Iversen, L. & Poulsen, K.B. (2001), Work environment factors, health, lifestyle and marital status as predictors of job change and early retirement in physically heavy occupations, *American Journal of Industrial Medicine*, Vol. 40(2), pp. 161–169.

Lingard, H. & Francis, V. (2005), Does work-family conflict mediate the relationship between job schedule demands and burnout in male construction professionals and managers? *Construction Management and Economics*, Vol. 23(7), pp. 733–745.

Lingard, H. & Turner, M. (2015), Improving the health of male, blue-collar construction workers: A social ecological perspective, *Construction Management & Economics*, Vol. 33(1), pp. 18–34.

Lingard, H. C., Yip, B., Rowlinson, S. & Kvan, T. (2007), The experience of burnout among future construction professionals: A cross-national study, *Construction Management and Economics*, Vol. 25(4), pp. 345–357.

Lynch, J. W., Kaplan, G. A. & Salonen, J. T. (1997), Why do poor people behave poorly? Variation in adult health behaviours and psychosocial characteristics by stages of the socioeconomic lifecourse, *Social Science & Medicine*, Vol. 44(6), pp. 809–819.

Ludewig, P. M. & Borstad, J. D. (2003), Effects of a home exercise program on shoulder pain and functional status in construction workers, *Occupational and Environmental Medicine*, Vol. 60(11), pp. 841–849.

MacDonald, L. A., Karasek, R. A., Punnett, L. & Scharf, T. (2001), Covariation between workplace physical and psychosocial stressors: Evidence and implications for occupational health research and prevention, *Ergonomics*, Vol. 44(7), pp. 696–718.

McDowell, I. & Newell, C. (1996), *Measuring health: A guide to rating scales and questionnaires* (2nd edn), Oxford University Press, New York.

McLaren, L. & Hawe, P. (2005), Ecological perspectives in health research, *Journal of Epidemiology and Community Health*, Vol. 59(1), pp. 6–14.

Maruish, M. E. (2011), *User's manual for the SF-36v2 health survey* (3rd edn), QualityMetric Incorporated, Lincoln, RI.

Meltzer, H., Griffiths, C., Brock, A., Rooney, C. & Jenkins, R. (2008), Patterns of suicide by occupation in England and Wales: 2001–2005, *British Journal of Psychiatry*, Vol. 193(1), pp. 73–76.

Noblet, A. & LaMontagne, A. D. (2006), The role of workplace health promotion in addressing job stress, *Health Promotion International*, Vol. 21(4), pp. 346–353.

Oude Hengel, K. M., Blatter, B. M., Joling, C. I., Van der Beek, A. J. & Bongers, P. M. (2012), Effectiveness of an intervention at construction worksites on work engagement, social support, physical workload and the need for recovery: Results from a cluster randomized controlled trial, BMC *Public Health*, Vol. 12, pp. 1008–1118.

Petersen, J. S. & Zwerling, C. (1998), Comparison of health outcomes among older construction and blue-collar employees in the United States, *American Journal of Industrial Medicine*, Vol. 34(3), pp. 280–287.

Pritchard, C. & Dixon, P. B. (2008), Reporting of skin cancer risks in the house-building industry: Alternative approaches to the analysis of categorical data, *Public Health*, Vol. 122(3), pp. 237–242.

Purdue, M. P., Järvholm, B., Bergdahl, I. A., Hayes, R. B. & Baris, D. (2006), Occupational exposures and head and neck cancers among Swedish construction workers, *Scandinavian Journal of Work, Environment & Health*, Vol. 32(4), pp. 270–275.

Roos, E., Lahelma, E. & Rahkonen, O. (2006), Work–family conflicts and drinking behaviours among employed women and men, *Drug and Alcohol Dependence*, Vol. 83(1), pp. 49–56.

Rushton, L., Hutchings, S. & Brown, T. (2008), The burden of cancer at work: Estimation as the first step to prevention, *Occupational and Environmental Medicine*, Vol. 65(12), pp. 789–800.

Safe Work Australia (SWA) (2016), *Construction industry profile* [Online]. Available at: www.safeworkaustralia.gov.au/sites/swa/statistics/industry/construction/pages/construction.

Sallis, J. F., Owen, N. & Fisher, E. B. (2008), Ecological models of health behaviour. In: K. Glanz, B. K. Rimer & K. Viswanath (eds), *Health behavior and health education: Theory, research, and practice* (4th edn), Jossey-Bass, San Francisco, pp. 465–86.

Schill, A. L. & Chosewood, L. C. (2013), The NIOSH Total Worker Health™ program: An overview, *Journal of Occupational and Environmental Medicine*, Vol. 55, pp. S8–S11.

Siebert, U., Rothenbacher, D., Daniel, U. & Brenner, H. (2001), Demonstration of the healthy worker survivor effect in a cohort of workers in the construction industry, *Occupational and Environmental Medicine*, Vol. 58(9), 774–779.

Sherratt, F. (2015), Legitimizing public health control on sites? A critical discourse analysis of the Responsibility Deal Construction Pledge, *Construction Management and Economics*, Vol. 33(5), pp. 444–452.

Siew, S. S., Kauppinen, T., Kyyrönen, P., Heikkilä, P. & Pukkala, E. (2012), Occupational exposure to wood dust and formaldehyde and risk of nasal, nasopharyngeal, and lung cancer among Finnish men, *Cancer Management and Research*, Vol. 4, pp. 223–232.

Snashall, D. (2005), Occupational health in the construction industry, *Scandinavian Journal of Work, Environment & Health*, Vol. 31(Suppl. 2), pp. 5–10.

Sorensen, G., Barbeau, E. M., Stoddard, A. M., Hunt, M. K., Goldman, R., Smith, A., Brennan, A. A. & Wallace, L. (2007), Tools for health: The efficacy of a tailored intervention targeted for construction laborers, *Cancer Causes Control*, Vol. 18(1), pp. 51–59.

Stattin, M. & Järvholm, B. (2005), Occupation, work environment and disability pension: A prospective study of construction workers, *Scandinavian Journal of Public Health*, Vol. 33(2), pp. 84–90.

Stocks, S. J., McNamee, R., Carder, M. & Agius, R. M. (2010), The incidence of medically reported work-related ill health in the UK construction industry, *Occupational and Environmental Medicine*, Vol. 67(8), pp. 574–576.

Stocks, S.J., Turner, S., McNamee, R., Carder, M., Hussey, L. & Agius, R.M. (2011), Occupational and work-related ill-health in UK construction workers, *Occupational Medicine*, Vol. 61(6), pp. 407–415.

Stokols, D., Grzwacz, J. G., Mcmahan, S. & Phillips, K. (2003), Increasing the health promotive capacity of human environments, *American Journal of Health Promotion*, Vol. 18, pp. 4–13.

Stokols, D., Allen, J. & Bellingham, R. L. (1996), The social ecology of health promotion: Implications for research and practice, *American Journal of Health Promotion*, Vol. 10(4), pp. 247–251.

SWA *see* Safe Work Australia.

Townsend, K., Loudoun, R. & Markwell, K. (2016), The role of line managers in creating and maintaining healthy work environments on project construction sites, *Construction Management and Economics*, Vol. 34(9), pp. 611–621.

Turner, M. & Lingard, H. (2016), Improving workers' health in project-based work: Job security considerations, *International Journal of Managing Projects in Business*, Vol. 9(3), pp. 606–623.

United Nations Organisation (UNO) (2013), *World population ageing*. Department of Economic and Social Affairs, Population Division, United Nations, New York.

UNO *see* United Nations Organisation.

Valero, E., Sivanathan, A, Bosché, F. & Abdel-Wahab, M. (2016), Musculoskeletal disorders in construction: A review and a novel system for activity tracking with body area network, *Applied Ergonomics*, Vol. 54, pp. 120–130.

Van der Hulst, M. (2003), Long work hours and health, *Scandinavian Journal of Work, Environment & Health*, Vol. 29(3), pp.171–188.

Van Steenbergen, E.F. & Ellemers, N. (2009), Is managing the work–family interface worthwhile? Benefits for employee health and performance, *Journal of Organizational Behavior*, Vol. 30(5), pp. 617–42.

Ware, J. E. (1999), SF-36 health survey in Maruish. In: E. Mark (ed.), (1999), *The use of psychological testing for treatment planning and outcomes assessment* (2nd edn), Lawrence Erlbaum Associates, Mahwah, NJ.

Welch, L.S. (2009), Improving work ability in construction workers – let's get to work, *Scandinavian Journal of Work Environment & Health*, Vol. 35(5), pp. 321–324.

WorkSafe Victoria (2016) Healthy workplace kit: a guide to implementing health and wellbeing programs at work [Online]. Available at: www.worksafe.vic.gov.au/wps/wcm/connect/f7093280439cdf6db37eb3145ee8dc5e/?a=17203 [Accessed 16 March 2017]

World Health Organization (2016), Asbestos: elimination of asbestos-related diseases – Fact Sheet [Online]. Available at: www.who.int/mediacentre/factsheets/fs343/en/ [Accessed 29 March 2017].

3 A model of how features of construction projects influence accident occurrence

Patrick Manu

The death of dogma is the birth of morality

(Immanuel Kant)

Background

With numerous injuries, deaths and work related illnesses reported in the construction industry of several countries (see HSE, 2014; U.S. Bureau of Labor Statistics, 2015), the need to tackle the health and safety (H&S) performance of the industry cannot be overstated. An important step towards tackling the H&S situation in the industry has been the emphasis on early planning of H&S in project delivery, which in the UK is mandatory under the Construction (Design and Management) Regulations (CDM Regulations). Central to the early planning of H&S is the need for construction project participants to pay attention to underlying accident causal factors that originate in the pre-construction stage (Haslam et al., 2005). In fact, studies have stressed that in order to prevent accidents on a long-term and sustainable basis (in other words, to have sustainable H&S improvement), there is a need to pay attention to these underlying accident causal factors (Haslam et al., 2005; Brace et al., 2009). This is confirmed by the fact that the pre-construction stage, from which underlying accident causes emanate, offers project participants the greatest opportunity to influence H&S on projects (Szymberski, 1997). Construction project features (CPFs) that constitute organisational attributes (e.g. procurement method), physical attributes (e.g. level of construction in terms of height/storey) and operational attributes (e.g. method of construction) of construction projects emanating from pre-construction decisions are among such underlying causes of construction accidents (Manu et al., 2012), and, as such, their H&S influence warrants attention.

This chapter presents an empirical inquiry into how CPFs influence accident occurrence. It commences by highlighting the accident causal influence of CPFs, and the gap in the existing literature in relation to how CPFs influence accident occurrence. This is followed by a review of accident causation models that enable a conceptual view of how CPFs influence accident occurrence. The research design employed in the study (i.e. a qualitative design) and the support-ing arguments are then presented. The research findings are then presented and discussed, the steps taken to validate the findings are explained, and, finally, conclusions are made.

The accident causal influence of CPFs

Despite the established significance of underlying accident causal factors to H&S (Haslam et al., 2005; Brace et al., 2009), there has been a dearth of empirical work with a focus on the accident causal role of CPFs. Even among the studies that have reported underlying causes of construction accidents, limited empirical focus has been accorded to the accident causal phenomenon of CPFs (see Suraji et al., 2001; Haslam et al., 2005; Brace et al., 2009; Cooke and Lingard, 2011). Although there have been reports on the accident implications of CPFs, such as the nature of the project, the method of construction used, the project duration, the level of construction, whether there is subcontracting, the design complex-ity, the type of site restriction, and the procurement system used (as summarised in Table 3.1), there are still aspects of this accident phenomenon that require empirical clarity. This is partly due to the inherent difficulties in examining underlying causes of accidents whose influence is slight, and which could thus go unnoticed (Haslam et al., 2005; Cooke and Lingard, 2011). An understanding of how CPFs influence accident occurrence is important if effective accident prevention measures are to be devised and implemented to mitigate the H&S impact of CPFs. Although the H&S literature is replete with accident causation models that attempt to explain how accidents occur, these models usually provide a generic view (Suraji et al., 2001), and often take a particular stand-point (e.g. human errors models, such as Hinze, 1996). Again, with the excep-tion of a few causation models (see Haslam et al., 2005), the models have also often focused on immediate/proximate causes of accidents (see Hinze, 1996). The existing models therefore do not specifically address how CPFs influence the occurrence of accidents.

Given the significance of underlying causes of accidents to construction H&S, empirical insight into how CPFs influence accident occurrence is thus warranted. A useful starting point is to consider the accident causation literature. The following section therefore reviews accident causation models/theories in rela-tion to the accident causal role of CPFs.

Accident causation models/theories

In reviewing the accident causation literature, a vital point is accident causation models and theories, which essentially attempt to explain how accidents occur in reality. Following the seminal work by Heinrich (1936), there have been subse-quent efforts to investigate how accidents occur, and these efforts have resulted in other accident causation models and theories. Prominent amongst these are energy transfer models, individual/human models/theories, and systems models.

Table 3.1 Example of literature sources highlighting the accident causal influence of CPFs

CPF\Literature Source	Mayhew and Quinlan (1997)	Egbu (1999)	Horbury and Hope (1999)	Gibb (1999, 2001)	Brabazon et al. (2000)	McKay et al. (2002)	Strategic Forum for Construction (2002)	Wright et al. (2003)	Perttula et al. (2003)	Hide et al. (2003)	Anumba et al. (2004)	Chua and Goh (2005)	Anumba et al. (2006)	Ankrah (2007)	Hughes and Ferrett (2008)	Brace et al. (2009)	HSE (2009)	Manu et al. (2013)
Nature of project		✓									✓		✓		✓		✓	
Method of construction				✓		✓	✓	✓	✓	✓								
Site restriction					✓					✓						✓		
Project duration	✓				✓					✓						✓		
Procurement system			✓		✓					✓						✓		
Design complexity					✓					✓						✓		
Level of construction												✓			✓		✓	
Subcontracting	✓		✓							✓				✓		✓		✓

Source: Adapted from Manu et al. (2012)

Energy transfer models consider causation of accidents as the release of uncontrolled energy from a source, where the energy is then conveyed through a path to the victim (Chua and Goh, 2004). Energy transfer models view accident occurrence as a one-dimensional phenomenon (from energy, through a path, to a receiver), despite the complexity and multi-causal nature of accidents (see Groeneweg, 1994). In terms of providing insight into how CPFs influence accident occurrence, the energy transfer models are thus unhelpful.

Individual/human models/theories emphasise the direct contribution of individuals to accident occurrence (Chua and Goh, 2004). They identify the causes and effects of errors/unsafe acts by individuals (usually frontline operatives), and they place emphasis on the psychological and behavioural aspects of humans (Chua and Goh, 2004). They do not explicitly facilitate continual improvement of workplace safety management systems, as they do not consider the role of organisation and management in the occurrence of accidents (Chua and Goh, 2004). Given the focus of individual theories of accident causation on individual factors (i.e. immediate accident causes), they are unhelpful in explaining how CPFs influence accident occurrence, as CPFs have an underlying causal influence.

Systems models of accident causation highlight the role of the organisation and its systems in the occurrence of accidents (Chua and Goh, 2004). They are concerned with the underlying causes of accidents (which are usually latent), the induced proximate causes, and the complex interactions between them. These models thus support the multi-causal nature of accidents, and they take a broader view of accident causation. Regarding systems models of accident causation, the constraint-response model and the ConCA model (Suraji et al., 2001; Haslam et al., 2005), in particular, could be useful in helping to explain how CPFs influence accident occurrence, as they highlight the causal influence of factors that are upstream of construction project procurement (e.g. decisions by the client, designers, and the project management team), and thereby provide an opportunity to address those factors timeously. The models thus drive home the message that accident prevention is not the sole responsibility of constructors, but also of other project participants, whose decisions dictate the way projects are physically executed. Being the result of pre-construction decisions by the client, designers, and the project management team, CPFs reflect the kind of underlying causal factors described by the constraint-response model as "distal factors", and by the ConCA model as "originating influences". Manu et al. (2012), drawing on the systems view of accident causation, particularly the constraint-response model and the ConCA model, proposed a conceptual model of how CPFs influence accident occurrence. Arguably, this model (illustrated in Figure 3.1) represents a useful step in shedding light on the mechanism by which CPFs influence accident occurrence.

Based on the generic systems view, that accidents are due to immediate causes triggered by underlying causes, the model proposes that CPFs influence accident occurrence through the inherent introduction of proximate causes of accidents into the construction phase of projects, to give rise to accidents.

Manu et al. (2012) therefore likened CPFs to Reason's (1990) resident pathogens, which are released by people who occupy a high position in the

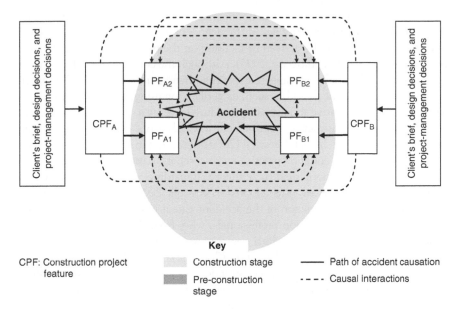

Figure 3.1 Conceptual model of the accident causal influence of CPFs (adapted from Manu et al., 2012)

decision-making structure of an organisation. After their release, CPFs (like resident pathogens), in turn, determine the nature, extent and existence of proximate accident causes on-site. Proximate causes are synonymous with the "proximal accident factors" in Suraji et al.'s (2001) constraint-response model and the "shaping factors" in Haslam et al.'s (2005) ConCA model. Also, based on the complexity and multi-causality of accidents, which the systems view of accident causation captures as complex inter-causal relationships, the model again proposes that the accident causal influence of CPFs is marked by inter-causal relationships between CPFs and the proximate causes. Such inter-causal relationships could manifest in the form of a CPF or a proximate factor eliminating, mitigating, or aggravating other proximate factors (Manu et al., 2012).

Despite the potential utility of the model in helping to explain how CPFs influence accident occurrence, its conceptual nature dictates that it first be verified empirically, to ascertain its credibility as a sound explanation of how CPFs influence accident occurrence. To this end, this chapter presents an empirical verification of how CPFs influence accident occurrence, through verification of the conceptual model.

Research method

Due to the interpretive focus of the study (the study focuses on how CPFs influence accident occurrence), a qualitative inquiry was undertaken, by the recommendation made by Fellows and Liu (2008). Seymour and Rooke (1995) have strongly advocated qualitative inquiry for construction management research. They explain that the utility of qualitative inquiry lies in the deeper understanding of the values and beliefs of others that can be derived by focusing on the points of view of individual practitioners (Seymour and Rooke, 1995). Although the conceptual model is a prior formulation, it essentially attempts to explain a phenomenon (i.e. how CPFs influence accident occurrence), and, as such, its verification using a qualitative inquiry is suitable. Use of qualitative inquiry as a procedure to verify a conceptualised view of a phenomenon in construction management research is not uncommon (see Tuuli and Rowlinson, 2010). In fact, Creswell (2009) notes that in other disciplines, such as health sciences, it is common practice for researchers to use qualitative inquiry to verify a prior formulation, such as a theory. Regarding strategy of inquiry, the phenomenological approach was adopted (Creswell, 2009). This was to enable exploration of the phenomenon of the accident causal influence of CPFs, through the experiences of construction professionals, who, from their vast experience on project sites, can relate to the H&S consequences of pre-construction decisions. Following the precedents of Haslam et al. (2005) and Choudhry and Fang (2008) regarding the use of interviews in construction accident causation studies, interviews, in particular semi-structured interviews, were adopted as the data-collection tool within the framework of the phenomenological strategy of inquiry. Following the application of this qualitative strategy, a credibility/validation check was also undertaken. Details of the check are presented later in the chapter.

The design of the interview

The proposed model has two key features: (1) the path of accident causation (i.e. CPFs, which introduce proximate factors, which, in turn, cause accidents); and (2) causal interactions (i.e. inter-causal relationships between CPFs and proximate factors, and also between proximate factors). In seeking verification of the model, the main objective of the interview was thus to explore the knowledge and experiences of experienced practitioners regarding the accident causal influence of CPFs, and the systems view of accident causation, with the intention of eliciting the two key features above. An interview schedule was used that consisted of a series of questions relating to: (1) how accidents are investigated within the practitioners' organisations; (2) the systems view of accident causation; (3) the H&S measures implemented by the practitioners' organisations on projects; (4) the accident causal influence of CPFs; and (5) the influence of pre-construction decisions by clients and project consultants (i.e. designers, and the project management team) in the occurrence of accidents. To support their views, the practitioners were also asked to relate accidents or near-miss events that they had witnessed on projects. No direct questions were asked regarding how CPFs influence accident occurrence, as conceptualised in the model. Rather, questions were asked indirectly, and the two key features were subsequently inferred from the responses. For instance, the practitioners were asked whether the H&S measures that they implement on projects are determined by or influenced by the features of the projects. The questions brought to the fore the accident causal influence of CPFs, and they made it possible to extract relevant issues relating to the path of accident causation, and the causal interactions between accident factors. The interviews were audio-taped, and, on average, they were approximately 60 minutes in duration.

Selection of participants

Using the UK Kompass online directory, 50 UK contractors were randomly selected, and were sent letters to solicit participation in the interviews. In the invitation, a request was made for a professional in a construction management role (e.g. an H&S manager, or a project manager, or a construction manager, or a site manager) to participate in the interview. H&S is a very sensitive subject in the UK, due to the legalities surrounding it, and, as a result, obtaining participation in H&S research is difficult (see Gibb et al., 2002). Given this state of affairs, it was deemed necessary to use industry contacts as well, to assist with obtaining participation in the interviews.

Compared with quantitative research, qualitative research involves fewer participants/cases, as the focus of qualitative research is not generalisation, but rather understanding of the meaning of phenomena, through collection and analysis of rich data (see Creswell, 2009; Choudhry and Fang, 2008; Mason, 2010). While determining sample size in qualitative research is difficult, some researchers have resorted to use of saturation point, which is the point reached when the data does not return new codes (see Mason, 2010). As qualitative data collection and analysis can run concurrently, it is possible to ascertain the point at which data returns no new codes, and where saturation has been reached. For instance, in Choudhry and Fang's (2008) construction accident causation study involving interviews, saturation was reached after seven interviews. Other studies in which saturation was reached at an early stage include Guest et al.'s (2006) study. Guest et al. (2006) concluded that for studies with high homogeneity among the population, a sample of even only six interviews may be adequate to enable development of meaningful themes and useful interpretations. In this study, saturation was reached after nine interviews. However, because prior arrangement had already been made with two other practitioners, an additional two interviews were conducted. The demographic information of the interviewees is given in Table 3.2. The participants are mainly contractor personnel in construction management roles, and they each have at least ten years of construction experience, which is an indication of adequate expertise in construction (Hallowell and Gambetese, 2010). Between them, the participants have an average of 26.27 years of construction experience.

Analysis

To aid the analysis, Creswell's (2009) guide for qualitative data analysis was used. The analysis followed five main steps: transcribing of the audio interviews (i.e.

Table 3.2 Respondents' demographic information

No.	Role of the participant	Years of experience in construction
1	Construction H&S consultant	30
2	H&S manager of a medium-sized national B&C*	10
3	Project manager of a large international B&C	34
4	Site manager of a medium-sized national B&C	20
5	H&S manager of a large international B&C	10
6	Senior site manager of a large national B&C	29
7	Civil engineer & director of a large international B&C	36
8	H&S manager of a large international B&C	20
9	Project manager of a large international B&C	42
10	Construction manager of a large international B&C	13
11	Project manager of a large international B&C	45

*B&C = building and civil engineering contractor. Annual turnover of medium-sized national B&C = circa £50 million. Annual turnover of large national B&C = circa £1 billion. Annual turnover of large international B&C = over £2 billion.

verbatim transcription); organising and preparing the transcripts; iterative rereading of the transcripts; coding of the transcripts; and generating themes. Creswell (2009) recommends that where a qualitative study seeks to verify a theory or prior formulation, coding should commence deductively (i.e. it should be informed by the literature), and it should then be complemented by inductive coding, as guided by emerging information from the transcripts. This recommendation was followed. The systematic iterative rereading and coding of the transcripts enabled attainment of a deep understanding of each interviewee's point of view, and hence extraction of issues related to the accident phenomenon under investigation.

Results and discussion

The findings are presented and discussed below.

Accident causation in construction

The analysis showed that the occurrence of construction accidents is viewed by the interviewees to be the result of immediate causes, which could be triggered by underlying causes. The responses of the interviewees also showed that while immediate causes tend to be relatively obvious, underlying causes are much more difficult to identify when investigating the causation of accidents. For instance, regarding the role of underlying causes in accident causation, and the difficulty in identifying such causes, interviewees made comments such as the following:

> Root causes are very important. One can have a fall from height. It might simply look like they slipped off a ladder, but then you start to question why the person slipped.
>
> (H&S manager)

Immediate causes are usually fairly obvious [...] finding root causes, in the first place, is the hardest bit.

(H&S manager)

In addition to the basic path of accident causation, i.e. from underlying causes through to immediate causes, the interviews also revealed that accident causation is a complex phenomenon characterised by interrelationships between causal factors. For instance, an interviewee portrayed the complex and multi-causal nature of accident causation as follows:

I think it is a very mixed picture. In some cases, you'd get causes that do influence each other [...] Definitely... things can interact and increase the chances of an accident taking place, without a shadow of doubt.

(H&S manager)

These findings are consistent with the systems view of accident causation and the systems models of construction accident causation by Suraji et al. (2001) and Haslam et al. (2005). Regarding accident investigations by the interviewees' organisations, it became increasingly evident that investigations do indeed try to trace the underlying causes of accidents. However, it also emerged from the interviews that the investigations focus on the underlying factors within the organisation's operations, and not on those factors that extend to the pre-construction stage. This is because, aside from the difficulty in establishing causality by those factors, it was felt that the contractors have no control, or very limited control, over decisions regarding those factors, and hence the need to rather focus on investigating factors that they can control.

How CPFs influence accident occurrence

Regarding the accident causal influence of CPFs, the analysis confirmed that the nature of the project, the method of construction used, the type of site restriction, the project duration, the procurement system used, the design complexity, the level of construction, and whether there is subcontracting have accident implications, as has previously been reported in the literature (see Table 3.1). Commenting on some of these features, one interviewee, for instance, emphasised that

[a] complex project brings more risk, a restricted site brings more risk, a tight duration brings more risk, and a high-rise also brings more risk, but you've got to manage those risks by putting in place the right measures to mitigate those risks.

(Project manager)

In addition to the above project features, another feature which emerged as having accident implications was restriction of site locality. This was drawn from a narrative of an accident, and also from elaborations given by interviewees on a closely related project feature, namely site restriction. The impact of site restriction is likely to occur as harm to a member, or members, of the public, as it involves working close to the public. Although it was acknowledged by the practitioners that the above project features have accident implications, an important view, which also emerged in the interviews, was that it comes down to how the risk associated with these project features is effectively managed right from the early stages of a project. This underscores the importance of effective H&S planning right from the early stages of project procurement (Szymberski, 1997), as well as the importance of having mechanisms in place such as the CDM Regulations.

Project features were considered to be underlying accident causal factors. An interviewee, for instance, referred to them as "*something that sits behind everything [...] they are underlying and quite deep underlying root causes*". The analysis further revealed that project features are associated with certain inherent H&S issues, which, as a result, make project features have the potential to influence the occurrence of accidents. Based on the interviews, the H&S issue associated with restriction of site locality is difficulty in traffic (pedestrian and vehicle) control around the site vicinity. The H&S issues are given in Table 3.3, and they are site-based. In the main, these H&S issues are similar to those that have been linked to CPFs in the existing construction and H&S literature (see McKay et al., 2002; Wright et al., 2003; Hide et al., 2003; Anumba et al., 2006; Baiden et al., 2006; Ankrah, 2007; Hughes and Ferrett, 2008; Brace et al., 2009; HSE, 2009). With CPFs being considered to be underlying/root causes with associated inherent H&S issues that are site-based, the interviews lend support to the first key feature of the proposed model, namely that CPFs introduce into the construction phase proximate causes of accidents, which, in turn, give rise to accidents.

Concerning the inter-causal relationships between project features, and proximate accident factors espoused by the conceptual model, the interviews also showed evidence of such relationships. For instance, some interviewees were of the view that while design-and-build procurement does not guarantee improved buildability, it does offer the opportunity to improve buildability of designs, due to contractor input in the design:

> Design-and-build gives you the opportunity to influence the design. I think the important thing with that is that a lot of construction companies may not actually realise they have that opportunity, and so even if they don't have novated designers, and they are their designers, they might simply say 'we are not the designers; we'll subcontract the design', without realising they have the opportunity to think about the designs, to review designs, and say 'well, hang on a minute; we'll never be able to build that, or that's going to be difficult to build safely, or expensive to build'.
>
> (H&S manager)

Table 3.3 CPFs and associated H&S issues

CPF	Associated H&S issue	An insightful comment on the H&S issue
Nature of project	Uncertainty of hazards	*"With new build, obviously you're starting from the ground, but with refurbishment you are working blindly. You don't know what is behind that plasterboard, do you?" (Site manager)*
Method of construction	Manual handling & mechanical handling	*"For example, the windows for this job arrived on site fully glazed [...] Lifting the windows by crane reduces manual handling risk, but it introduces risk associated with operating a crane." (Site manager)*
Site restriction	Congestion	*"The separation of plant, workers, materials, and vehicles is more difficult on restricted site, and so it's more dangerous to work on restricted sites." (Site manager)*
Project duration	Time pressure	*"When the duration is tight, the workers are under pressure, and when they are under pressure, they'll cut corners if you allow them. As soon as time pressure is introduced, accidents can occur." (Site manager)*
Procurement system	Difficulty in collaborative working	*"I'll say some of the collaborative early contractor involvement-type of procurement helps us to think through problems and things in more detail." (Civil engineer & director)* *"It does pay dividends to be working with the design team months in advance before starting on site. And, of course, through that you build a good relationship with the design team, as well as some trust." (Project manager)*
Design complexity	Difficulty in building (i.e. buildability)	*"I think one of the key issues is for designers to understand that it might look very good on paper, but someone has to deliver the design in operational terms." (Construction H&S consultant)*
Level of construction	Working at height	*"The level of construction could influence accident occurrence because of working at height. I feel more confident, and everybody feels more confident, the lower they are working." (Site manager)*
Subcontracting	Fragmentation of workforce	*"One of the big challenges for the industry is the subcontract culture [...] it is not unheard of for a team to turn up on site and they don't even know who we are, because they've been contracted by somebody who has been contracted by somebody." (H&S manager)*

Haslam et al. (2005) in their study similarly reported this perception about design-and-build. This means that design-and-build offers the opportunity to reduce the difficulty in construction (i.e. the site-based H&S issue/proximate cause) associated with design complexity, and this provides an example of the possible inter-causal relationships between some CPFs, and the proximate accident factors introduced by other CPFs. Regarding inter-causal relationships between the various site-based H&S issues (i.e. the proximate causes), the analysis did not reveal specific examples. Nonetheless, from the interviewees' general acknowledgement of possible inter-causal relationships between accident factors, the possibility of there being inter-causal relationships between the site-based H&S issues in the process of accident occurrence cannot be rejected. The interviews also lend support to the second key feature of the model.

Credibility/validation check

Demonstrating credibility in qualitative research is important in establishing confidence in the findings, and in this regard, some checks have been suggested, e.g. verbatim transcription of interviews (see Creswell, 2009). In this study, a further check that was applied was member checking/respondent validation (Silverman, 2006; Creswell, 2009). This involved a follow-up questionnaire survey of construction professionals which investigated the H&S impact of CPFs. An aspect of the survey was used to validate the interview findings, particularly the finding that CPFs are inherently associated with H&S issues (proximate accident causes) that are introduced into the construction phase. The survey yielded 184 valid responses (out of 1,000 administered questionnaires) from UK construction professionals in construction management roles (e.g. H&S manager, project manager, site manager, and construction manager). Together, the professionals have a respectable 16.30 years of experience in their respective roles, and an average of 24.31 years of experience in construction. Approximately 80% of them have over ten years of experience in construction. Approximately 70% of the respondents are members of at least one industrial professional body (e.g. the Institution of Civil Engineers, the Institution of Occupational Safety and Health, the Chartered Institute of Building, the International Institute of Risk and Safety Management, or the Royal Institution of Chartered Surveyors). The demographic information shows that the respondents are adequately experienced in the management of construction and H&S (see Hallowell and Gambetese, 2010).

Relying on their broad construction experience, the respondents were asked to rate the extent to which the H&S issues in Table 3.3 are common/prevalent within their associated CPFs. A five-point scale (0 = not at all, 1 = low, 2 = moderate, 3 = high, 4 = very high) was used. The results are shown below in Table 3.4. Table 3.4 shows the mean ratings of the respondents. For the mean ratings to be interpreted with confidence, evidence of agreement amongst the respondents is important. The table therefore also shows single-item inter-rater agreement indices (r_{WG}), which test for consensus amongst the respondents

Table 3.4 Extent to which inherent health and safety issues (proximate causes of accidents) are common/prevalent within CPFs

Construction project feature (CPF)	Extent to which proximate cause of accident is common/prevalent within CPF	Med.	Mode	Mean	Std dev.	r_{WG}*	Overall assessment High (3)	Overall assessment Moderate (2)	Overall assessment Low (1)
Nature of project (New build, refurbishment, and demolition)	Uncertainty of hazards within refurbishment	3.00	3.00	2.7714	.93629	0.56	✓		
	Uncertainty of hazards within demolition	3.00	3.00	2.9324	.94803	0.55	✓		
	Uncertainty of hazards within new work	2.00	1.00	1.6246	.73580	0.73			✓
Level of construction (High-level construction, and low-level construction)	Working at height within high-level construction (i.e. multi-level construction)	3.00	3.00	3.1832	.85076	0.64	✓		
	Working at height within low-level construction (i.e. single-level construction)	2.00	2.00	1.9756	.89674	0.60		✓	
Subcontracting (Multi-layer and single-layer subcontracting)	Fragmentation of workforce within single-layer subcontracting	2.00	2.00	1.7728	.72886	0.73		✓	
	Fragmentation of workforce within multi-layer subcontracting	3.00	3.00	2.7273	.80241	0.68	✓		
Procurement system (Traditional, design-and-build, partnering, and management contracting)	Fragmentation of project team within traditional procurement	2.00	2.00	1.8553	.74317	0.72		✓	
	Fragmentation of project team within design-and-build procurement	2.00	2.00	1.8109	.73153	0.73		✓	
	Fragmentation of project team within partnering procurement	2.00	2.00	1.8198	.77830	0.70		✓	
	Fragmentation of project team within management contracting	2.00	2.00	2.0225	.70703	0.75		✓	
Method of construction (Pre-assembly, and traditional method)	Manual handling within pre-assembly construction	2.00	2.00	1.7465	.77017	0.70		✓	
	Manual handling within traditional construction	3.00	3.00	2.6614	.69753	0.76	✓		
	Mechanical handling within pre-assembly construction	3.00	3.00	2.4021	.91827	0.58		✓	
	Mechanical handling within traditional construction	2.00	2.00	2.3238	.72411	0.74		✓	

(Continued)

Table 3.4 Extent to which inherent health and safety issues (proximate causes of accidents) are common/prevalent within CPFs (Continued)

Construction project feature (CPF)	Extent to which proximate cause of accident is common/prevalent within CPF	Med.	Mode	Mean	Std dev.	r_{wg}*	Overall assessment		
							High (3)	Moderate (2)	Low (1)
Project duration (Adequate duration, and tight duration)	Time pressure within tight project duration	3.00	3.00	3.1322	.67841	0.77	✓		
	Time pressure within adequate project duration	2.00	2.00	1.7843	.71232	0.75		✓	
Site restriction (Restricted site, and unrestricted site)	Site congestion within restricted site (i.e. where footprint of facility covers most of the site area)	3.00	3.00	3.0472	.71876	0.74	✓		
	Site congestion within unrestricted site (i.e. where footprint of facility covers a smaller portion of the site area)	2.00	2.00	1.5992	.68854	0.76		✓	
Design complexity (Complex design, and simple design)	Difficulty in constructing within complex design (i.e. design with intricate aesthetic qualities)	3.00	3.00	2.8957	.76707	0.70	✓		
	Difficulty in constructing within simple design (i.e. design with simple aesthetic qualities)	1.00	1.00	1.4367	.65512	0.79			✓
Restriction of site locality (Restricted site locality, and unrestricted site locality)	Difficulty in traffic control around site vicinity within restricted site locality (e.g. city-centre location)	3.00	3.00	3.0732	.69869	0.76	✓		
	Difficulty in traffic control around site vicinity within unrestricted site locality (e.g. outer-city location)	2.00	1.00	1.6104	.75104	0.72		✓	

* r_{wg} indices > 0.14 are significant at $p < 0.001$, based on 10,000 simulation runs, a group size of 184, and five response options (i.e. a five-point scale) (see Cohen et al. (2001)).

(James et al., 1984). The estimated r_{WG} indices are evidence of significant consensus amongst the respondents. Approximation of the mean ratings to their nearest scale points confirms that the respective H&S issues are inherently associated with their respective CPF, as none of the mean ratings approximates to the point of zero, which would mean that the H&S issue in question is not at all common/prevalent within the CPF. On the contrary, the H&S issues are common/prevalent within their associated CPFs to varying extents, ranging from low to high.

Silverman (2006) argues that where member checking/respondent validation is applied, where participants verify the findings of the research, it generates more confidence in the credibility of the findings. The convergence between the results of the validation check and the findings drawn from the interviews lends confidence to the credibility of the interview findings, and hence the soundness of the proposed model of how CPFs influence accident occurrence.

Summary

While it is now evident that CPFs have a causal influence in accident occurrence, through the inherent introduction of proximate accident factors, the study presented in this chapter has also indicated that the accident causal influence of CPFs could be known by causal interactions between CPFs and proximate accident factors. This causal interaction can reduce or increase the presence of proximate accident factors. These findings lend empirical support and credence to the proposed model, as shown in Figure 3.1, of the mechanism by which CPFs influence accident occurrence. Contributing to earlier construction accident causation models, the model places the spotlight on features of construction projects, amongst other underlying causal factors in construction accidents, and it explains how various CPFs acting collectively on a project contribute to the occurrence of accidents. As accident causation models are useful in devising and implementing accident prevention measures, the model could similarly be useful to pre-construction project participants during the early stages of projects. The model could serve as evidence-based justification for encouraging pre-construction project participants to devise and implement accident prevention measures which can remove or 'block' the release of proximate factors introduced by CPFs. Also, knowledge of the existence of potential causal interactions between CPFs and proximate accident causes could be a basis for pre-construction project participants to carefully consider selection or avoidance of certain combinations of CPFs, due to the effects that some CPFs could have on the proximate factors associated with other CPFs. This could be very useful, particularly where due to certain project conception constraints, some CPFs are inevitable. Also, in terms of investigating accident causation on projects, the model could serve as a root-cause analysis tool in helping to trace or probe the potential contribution of CPFs, and hence the contribution of pre-construction project participants in accident occurrence. For instance, if an accident investigation reveals involvement of any of the H&S issues/proximate causes, further probing as to why those causes were involved could point in the direction of certain CPFs, and hence the contribution of certain pre-construction project participants.

The study, however, has a limitation, which makes further research desirable. While this research has shown that there could be inter-causal relationships between CPFs and the proximate accident factors that they introduce, the study does not provide insight into the extent to which various CPFs could increase or decrease the prevalence of proximate factors, e.g. can a procurement system increase or decrease the extent of time pressure introduced by tight project duration, and if so, to what extent can it do this? Further research in this direction would provide additional insight, which would strengthen the utility of the model for pre-construction H&S risk mitigation.

References

Ankrah, N.A. (2007). *An investigation into the impact of culture on construction project performance*. PhD thesis. University of Wolverhampton.

Anumba, C., Egbu, C. and Kashyap, M. (2006). *Avoiding structural collapses in refurbishment: A decision support system*. Suffolk: HSE Books.

Anumba, C., Marino, B., Gottfried, A. and Egbu, C. (2004). *Health and safety in refurbishment involving demolition and structural instability*. Suffolk: HSE Books.

Baiden, B.K., Price, A.D.F. and Dainty, A.R.J. (2006). The extent of team integration within construction projects. *International Journal of Project Management*, 24(1): 13–23.

Brabazon, P., Tipping, A. and Jones, J. (2000). *Construction health and safety for the new Millennium*. Suffolk: HSE Books.

Brace, C., Gibb, A., Pendlebury, M. and Bust, P. (2009). *Phase 2 Report: Health and safety in the construction industry: Underlying causes of construction fatal accidents: External research*. Norwich: Her Majesty's Stationery Office.

Choudhry, R.M. and Fang, D. (2008). Why operatives engage in unsafe work behavior: Investigating factors on construction sites. *Safety Science*, 46(4): 566–584.

Chua, D.K.H. and Goh, Y.M. (2004). Incident causation model for improving feedback of safety knowledge. *Journal of Construction Engineering and Management*, 130(4): 542–551.

Chua, D.K.H. and Goh, Y.M. (2005). Poisson model of construction incident occurrence. *Journal of Construction Engineering and Management*, 131(6): 715–722.

Cohen, A., Doveh, E. and Eick, U. (2001). Statistical properties of the $r_{WG}(J)$ index of agreement. *Psychological Methods*, 6(3): 297–310.

Cooke, T. and Lingard, H. (2011). "A retrospective analysis of work-related deaths in the Australian construction industry", in 27th Annual ARCOM Conference on 5–7 September in Bristol, UK, Association of Researchers in Construction Management, pp. 279–288.

Creswell, J.W. (2009). *Research design: Qualitative, quantitative, and mixed method approaches*. 3rd edn. Thousand Oaks, CA: Sage.

Egbu, C.O. (1999). Skills, knowledge and competencies for managing construction refurbishment works. *Construction Management and Economics*, 17(1): 29–43.

Fellows, R. and Liu, A. (2008). *Research methods for construction*. West Sussex: Blackwell.

Gibb, A.G., Haslam, R.A., Gyi, D.E., Hide, S., Hastings, S. and Duff, R. (2002). ConCA: Preliminary results from a study of accident causality, in Triennial Conference CIB W099, May, University of Hong Kong, Hong Kong.

Gibb, A.G.F. (1999). *Principles in off-site fabrication*. Caithness, UK: Whittles.

Gibb, A.G.F. (2001). Standardization and pre-assembly: Distinguishing myth from reality using case study research. *Construction Management and Economics*, 19(3): 307–315.

Groeneweg, J. (1994). *Controlling the controllable: The management of safety.* 2nd revised edn. Leiden: DSWO Press.

Guest, G., Bunce, A. and Johnson, L. (2006). How many interviews are enough? An experiment with data saturation and variability. *Field Methods*, 18(1): 59–82.

Hallowell, M.R. and Gambatese, J.A. (2010). Qualitative research: Application of the Delphi method to CEM research. *Journal of Construction Engineering and Management*, 136(1): 99–107.

Haslam, R.A., Hide, S.A., Gibb, A.G.F., Gyi, D.E., Pavitt, T., Atkinson, S. and Duff, A.R. (2005). Contributing factors in construction accidents. *Applied Ergonomics*, 36(4): 401–415.

Heinrich, H.W. (1936). *Industrial accident prevention.* New York: McGraw-Hill.

Hide, S., Atkinson, S., Pavitt, T., Haslam, R., Gibb, A., Gyi, D., Duff, R. and Suraji, A. (2003). *Causal factors in construction accidents.* Suffolk: HSE Books.

Hinze, J. (1996). "The distraction theory of accident causation", in Implementation of safety and health on construction sites: Proceedings of the First International Conference of CIB Working Commission W99: Safety and Health in Construction, 4–7 September, in Lisbon, A.A. Balkema, Rotterdam, pp. 375–384.

Horbury, C. and Hope, C. (1999). "The impact of procurement and contracting practices on health and safety: A review of literature", Report RAS/99/02, Buxton, HSL.

HSE (2009). "Construction intelligence report: Analysis of construction injury and ill-health intelligence". Health and Safety Executive, Available online at http://s3. amazonaws.com/zanran_storage/www.hse.gov.uk/ContentPages/25543732.pdf (Accessed 1 January 2011).

HSE (2014). "Health and safety in construction in Great Britain, 2014: Work-related injuries and ill health". Health and Safety Executive. Available online at www.hse.gov. uk/statistics/industry/construction/construction.pdf (Accessed 22 June 2015).

Hughes, P. and Ferrett, E. (2008). *Introduction to health and safety in construction.* 3rd edn. Oxford: Elsevier.

James, L.R., Demaree, R.G. and Wolf, G. (1984). Estimating within-group interrater reliability with and without response bias. *Journal of Applied Psychology*, 69(1): 85–98.

Manu, P., Ankrah, N., Proverbs, D. and Suresh, S. (2012). Investigating the multi-causal and complex nature of the accident causal influence of construction project features. *Accident Analysis & Prevention*, 48: 126–133.

Manu, P., Ankrah, N., Proverbs, D. and Suresh, S. (2013). Mitigating the health and safety influence of subcontracting in construction: The approach of main contractors. *International Journal of Project Management*, 31(7): 1017–1026.

Mason, M. (2010). Sample size and saturation in PhD studies using qualitative interviews. *Forum: Qualitative Social Research*, 11(3), Art. 8.

Mayhew, C. and Quinlan, M. (1997). Subcontracting and occupational health and safety in the residential building industry. *Industrial Relations Journal*, 28(3): 192–205.

McKay, L.J., Gibb, A.G.F., Haslam, R. and Pendlebury, M. (2002). "Implications for the effect of standardization and pre-assembly on health, safety and accident causality: Preliminary results", in 18th Annual ARCOM Conference, 2-4 September, University of Northumbria. Association of Researchers in Construction Management.

Perttula, P., Merjama, J., Kiurula, M. and Laitinen, H. (2003). Accidents in materials handling at construction sites. *Construction Management and Economics*, 21(7): 729–736.

Reason, J. (1990). *Human error.* Cambridge, UK: Cambridge University Press.

Seymour, D. and Rooke, J. (1995). The culture of the industry and the culture of research. *Construction Management and Economics*, 13(6): 511–523.

Silverman, D. (2006). *Interpreting qualitative data: Methods for analyzing talk, text, and interaction*. London: Sage.

Strategic Forum for Construction (2002). "Accelerating change: A report by the Strategic Forum for Construction, chaired by Sir John Egan". London: Rethinking Construction.

Suraji, A., Duff, A.R. and Peckitt, S.J. (2001). Development of causal model of construction accident causation. *Journal of Construction Engineering and Management*, 127(4): 337–344.

Szymberski, R.T. (1997). Construction project safety planning. *TAPPI Journal*, 80(11): 69–74.

Tuuli, M.M. and Rowlinson, S. (2010). What empowers individuals and teams in project settings? A critical incident analysis. *Engineering, Construction and Architectural Management*, 17(1): 9–20.

U.S. Bureau of Labor Statistics (2015). National census of fatal occupational injuries in 2014. U.S. Bureau of Labor Statistics. Available online at www.bls.gov/news.release/pdf/cfoi.pdf (Accessed 9 October 2016).

Wright, M., Bendig, M., Pavitt, T. and Gibb, A. (2003). The case for CDM: Better safer design – a pilot study. Suffolk: HSE Books.

4 Profiling the health and safety risk associated with construction projects at the pre-construction stage

Patrick Manu

Science is organized knowledge. Wisdom is organized life.–

(Immanuel Kant)

Background

The construction industry is widely noted to be a dangerous industry in respect of the health and safety (H&S) of workers (ILO, 2005). Despite years of gradual improvements in construction H&S in some countries, the industry continues to record numerous deaths, injuries, and ill health among workers in many countries (see Department of Occupational Safety and Health, 2015; HSE, 2015a; Ministry of Manpower, 2015). As a result, efforts continue to be made to enhance the H&S situation of the industry. Among the efforts is the emphasis on the early planning of H&S in project procurement. For instance, in the UK there are the Construction Design and Management (CDM) Regulations 2015, which seek to integrate H&S into construction project delivery right from the early design stage (HSE, 2015b), by placing legal duties on key project participants, including clients and designers. Similar regulations also exist in countries such as Singapore and Australia (e.g. the Workplace Safety and Health (Design for Safety) Regulations 2015 of Singapore, and the Work Health and Safety Acts and Regulations of several jurisdictions in Australia). Central to the early planning of H&S is the need for project participants to pay attention to underlying causes of accidents which emanate from the pre-construction phase of project delivery.

Construction project features (CPFs), which are organisational, physical and operational attributes of construction projects, are among underlying accident causal factors, and they include the nature of the project, the method of construction used, the project duration, the level of construction, whether there is subcontracting, the design complexity, the type of site restriction, and the procurement system used (see Mayhew and Quinlan, 1997; Egbu, 1999; Horbury and Hope, 1999; Brabazon et al., 2000; McKay et al., 2002; Strategic Forum for Construction, 2002; Wright et al., 2003; Perttula et al., 2003; Hide et al., 2003; Chua and Goh, 2005; Anumba et al., 2006; Ankrah, 2007; Hughes and Ferrett, 2008; Brace et al., 2009; HSE, 2009; Manu et al., 2013). These features mainly emanate from pre-construction decisions by the client, designers, and the project managers/planners.

At the pre-construction stage of construction projects, it is important for project participants to have an understanding of the H&S risk associated with a project so that the appropriate mitigation measures can be introduced before construction. This understanding is also necessary for clients and their representatives to be able to appoint project team members (e.g. contractors) who have the competence to manage the H&S risk in a project. In the UK, for example, the Construction Design and Management (CDM) Regulations 2015 recommend that H&S risk measures be proportionate to the risks involved in a project (HSE, 2015b). The CDM 2015 also requires project participants to have the knowledge, skills, and experience (if they are individuals) and the organisational capability (if they are an organisation) to manage the H&S risks in a project (HSE, 2015b). CPFs, being underlying accident causal factors and the result of pre-construction decisions, can be informative in profiling the H&S risk associated with a project at the pre-construction stage. This chapter presents a pathway for ascertaining the H&S risk profile of a project, through an evaluation of the H&S risk associated with the features of the project. In the sections that follow, a review of approaches for assessing H&S risk is considered, to determine a suitable method for evaluating the H&S risk associated with CPFs. This is followed by the evaluation of the H&S risk associated with CPFs, and the development of a simple pre-construction-stage H&S risk-profiling tool (nicknamed "CRiMT").

Health and safety risk evaluation

Regarding H&S, several definitions have been used for risk, including: *"the combination of the likelihood of an occurrence of a hazardous event or exposure(s) and the severity of injury or ill health that can be caused by the event or exposure(s)"* (British Standard Institute (BSI), 2008); *"the likelihood of a substance, activity or process to cause harm"* (Hughes and Ferrett, 2008). According to the HSE (2001, 2006), the risk is the likelihood that harm will occur. By conceiving of H&S risk associated with a CPF as the probability of that CPF causing harm (e.g. injury or illness), the definitions by the HSE (2001, 2006) and Hughes and Ferrett (2008) can be instructive in understanding the H&S risk associated with CPFs. It has been noted that the terms "risk" and "hazard" are often used interchangeably, even though these two terms do not mean the same thing (HSE, 2001). A hazard is the intrinsic potential of something to cause harm (HSE, 2001), and the HSE (2001) further relates hazard to risk, in this manner: *"risk is the chance that someone or something that is valued will be adversely affected in a stipulated way by a hazard."* Focusing on the H&S of people, and about hazard, the risk is thus the likelihood that someone will be harmed (i.e. adversely affected) by a hazard (i.e. the potential of something to cause harm).

The methods for evaluating H&S risk have been categorised mainly as qualitative and quantitative risk evaluation (see BSI, 2008; Pinto et al., 2011). Popular among the qualitative methods is the checklist/questionnaire method (see Pinto et al., 2011). Quantitative methods include failure modes and effects analysis, and hazard and operability studies (see BSI, 2008). There are also semi-quantitative, or qualitative-quantitative, methods, which quantify qualitative risk information or use qualitative risk information with corresponding numeric scores (see Aven, 2008; Sachs and Tiong, 2009; WHO and FAO, 2009). A popular method of semi-quantitative risk evaluation is the risk combination matrix (cf. Risk & Policy Analyst Ltd, 1999; WHO and FAO, 2009), which some authors (e.g. Hughes and Ferrett, 2008) prefer to classify as a quantitative method.

The literature provides examples of applications of the different types of H&S risk evaluation methods. For instance, Kariuki and Löwe (2007) developed a qualitative risk evaluation method that systematically identifies the human error in process design and the human factors that influence the production and propagation of human error. The New York State Division of Industrial Safety Services used a quantitative method that correlates the degree of risk of various construction activities and workmen's compensation insurance rates (Knab, 1978). The technique adopted by the New York State Division of Industrial Safety Services was modified by Knab (1978), who developed a model that determines a risk score for various workmen's compensation classifications. Sacks et al. (2009) developed the construction hazard assessment with spatial and temporal exposure (CHASTE) method, which numerically estimates safety risk levels, based on the following: the probability of exposure in space and time; an estimate of the likelihood of a loss-of-control event occurring per worker in a crew; the expected severity of the result of a potential accident; and the numbers of employees in the team. Regarding semi-quantitative risk evaluation, Croner (in Risk & Policy Analyst Ltd, 1999) proposed a task-based method for rating risk, using a risk matrix which combines scores for severity and likelihood of hazard. The product of these scores (i.e. the result obtained by multiplying them together) provides a risk rating ranging from 1 (chances of death very unlikely) to 100 (certain/imminent multiple deaths). The matrix classifies risk as trivial, adequately controlled, or not properly controlled, based on the need for further action. Hughes and Ferrett (2008) also proposed a risk matrix, which estimates risk by combining severity of harm (measured by a 3-point scale: slight (1), severe (2), and major (3)) and likelihood of harm (measured by a 3-point scale: low (1), medium (2), and high (3)). The various combinations are assigned a degree of risk ranging from low risk (for combinations 1–2), to medium risk (for combinations 3–4), to high risk (for combinations 6–9). Jannadi and Almishari (2003) also developed a risk assessor model, which determines a risk score for an activity, and a justification factor for a proposed remedy. The risk score is estimated based on severity, exposure, and probability, which are all determined using qualitative scales (e.g. minor cuts, for severity; occasionally, for the frequency of exposure; and likely, for probability), and are tied to quantitative scores or ratings.

The above demonstrates the variety of risk evaluation methods applied in H&S. As noted by several authors (e.g. Smith et al., 2006; Pinto et al., 2011), generally the various methods are valid depending on the context of the application. However, these methods have their strengths and weaknesses, which should be taken into account when selecting a risk evaluation method.

Although qualitative risk assessment is easy to use, it is subjective, and thus makes it difficult for a third party to understand the basis or rationale for the evaluation (WHO and FAO, 2009). Also, qualitative risk assessment mainly prioritises identified hazards for further analysis, and, as such, it does not necessarily give an indication of the likelihood of occurrence of harm (i.e. risk) (WHO and FAO, 2009). A qualitative risk evaluation will thus not be suitable for evaluating the H&S risk associated with CPFs. With quantitative risk evaluation, the major challenge is the availability or completeness of any historical and numerical data needed for the evaluation (BSI, 2008; Sachs and Tiong, 2009). Previous construction accident causation studies in the UK and elsewhere have noted the limitations of accident records, especially where it comes to analysing root accident causal factors which are upstream of the project procurement process (see Suraji, 2001; BOMEL Limited et al., 2006; Cooke and Lingard, 2011). Given that CPFs fall into this category of accident causal factors, a purely quantitative risk evaluation will also not be a suitable option.

Semi-quantitative risk evaluation provides an intermediate level of qualitative risk evaluation and quantitative risk assessment. It offers a more consistent and rigorous approach to evaluating and comparing risks than does a qualitative risk assessment, and it avoids some of the greater ambiguities that a qualitative risk assessment may produce (WHO and FAO, 2009). It does not require the same mathematical skill as quantitative risk evaluation, nor does it require the same amount of data, which means that it can be applied to risks where precise data are missing or unavailable. Semi-quantitative risk evaluation thus appears to be a more suitable approach for assessing the H&S risk associated with CPFs. Semi-quantitative risk assessment is, however, not without its weaknesses. The resulting risk scores are placed into usually quite broad sets of categories (e.g. risk score 0–3 = low risk; risk score 4–7 = medium risk; risk score 8–10 = high risk). This weakness can, however, be overcome if the categories are carefully constructed/defined (WHO and FAO, 2009). As with purely quantitative risk evaluation, a key aspect of semi-quantitative risk evaluation is the risk expression based on which risk is evaluated. The risk expression indicates the risk-determining variables, and, again, several risk expressions exist in the H&S literature. Among the common ones are:

- Risk = likelihood × severity (Risk & Policy Analyst Ltd, 1999; Hughes and Ferrett, 2008);
- Risk = probability × severity x exposure (Jannadi and Almishari, 2003); and
- Risk = probability × severity x frequency (Risk & Policy Analyst Ltd, 1999).

Another common, and hence widely used, the expression is *risk = hazard × exposure* (Chicken and Posner, 1998; Duffus and Worth, 2001; Canadian Centre for Occupational Health and Safety, 2008). By this expression, risk (i.e. the likelihood of occurrence of harm) is a function of hazard (i.e. the potential of a thing to cause harm) and exposure to the hazard. This expression is supported by the argument that unless there is the hazard, there cannot be the risk (HSE, 2000). The role of hazard in determining risk is emphasised by the following statement: "*risk is the chance that someone or something that is valued will be adversely affected in a stipulated way by a hazard*" (HSE, 2001). The expression also shows that hazard alone does not determine risk, but it does so through exposure of people to the hazard. Duffus and Worth (2001) support this with the argument that unless there is exposure to a hazard, there will be no risk, regardless of the degree of the hazard. Hazard, being the potential of a thing to cause harm, coupled with exposure thus determines risk.

Considering the accident causal influence of CPFs, their potential to influence accident occurrence can be taken as their potential to cause harm to people, as accidents ultimately result in harm (e.g. injury and ill health). The H&S risk associated with CPFs (i.e. the likelihood of occurrence of accidents/harm) can thus be considered regarding the expression *risk = hazard × exposure*, where "hazard" (i.e. the potential to cause harm) is taken as the potential of CPFs to influence accident occurrence. "Exposure" is the extent to which people or objects are subjected to a hazard (Canadian Centre for Occupational Health and Safety, 2008), and it can be assessed in various forms, such as duration, frequency, concentration, inhalation, and contact (see Duffus and Worth, 2001; Canadian Centre for Occupational Health and Safety, 2008). Referring to semi-quantitative risk evaluation as a viable approach for evaluating the H&S risk associated with CPFs, a risk matrix can be used, given its wide use, and also considering that the mathematical risk expression *risk = hazard × exposure* can easily be applied in a two-dimensional matrix format when adapted for the context of H&S risk associated with CPFs. Regarding the H&S risk associated with CPFs, the above expression can thus be rewritten as follows:

H&S risk associated with a CPF = the potential of the
CPF to influence accident occurrence × exposure of the workforce.   (1)

To evaluate the H&S risk associated with CPFs, the next section explains how the inputs of the expression were determined and subsequently applied.

Research methods

The methods/steps applied are explained in the following sections.

Determining the potential of CPFs to influence accident occurrence

To enable measurement of the potential of CPFs (i.e. the nature of the project, the method of construction used, the project duration, the level of construction, whether there is subcontracting, the design complexity, the type of site restriction, and the procurement system used) to influence accident occurrence, a quantitative approach, in particular a survey, was adopted. Contractor personnel, including site managers, construction managers, H&S managers, and project managers, commonly work on construction project sites in management roles, and from their extensive industrial experience in construction they are likely to be aware of the above CPFs, as well as their impact on H&S in projects. These professionals were therefore targeted in a survey, to enable a generic assessment of CPFs' potential to influence accident occurrence. The reported difficulty in studying underlying causes of accidents using accident records/cases (see Suraji et al., 2001; BOMEL Limited et al., 2006; Cooke and Lingard, 2011) also supported use of an alternative research approach, in this case a survey, to determine the potential of CPFs to influence accident occurrence.

A questionnaire was designed to measure the degree of potential of CPFs to influence accident occurrence, using a 5-point Likert-type scale (0 = none, 1 = low, 2 = moderate, 3 = high, 4 = very high), similar to that used in a previous accident causation study (see Hide, 2003). To reach the above groups of contractor personnel, a survey was conducted on a sample of contractors randomly drawn from the UK Kompass online directory. A total of 1,000 questionnaires were administered to contractors, requesting the participation of a professional in a construction management role. The survey yielded 184 valid responses, equating to a response rate of 18.4%. The respondents' roles are: project manager (7.61%), construction manager (15.76%), site manager (2.17%), H&S manager (56.52%), and other construction roles (e.g. managing director, construction director, operations manager, and integrated manager (health, safety, environment, and quality)) (17.94%). Between them, the professionals have an average of 16.30 years of experience in their respective roles, and an average of 24.31 years of experience in construction. Approximately 80% of them have over ten years of experience in construction. Approximately 70% of the respondents are members of at least one industrial, professional body (e.g. the Institution of Civil Engineers, the Institution of Occupational Safety and Health, the Chartered Institute of Building, the International Institute of Risk and Safety Management, the Institution of Engineers of Ireland, the Institution of Royal Engineers, and the Royal Institution of Chartered Surveyors). The survey results presented in Table 4.1 show the minimum, the maximum, the mean score, and the standard deviation regarding the potential of CPFs to influence accident occurrence. The table also shows single-item inter-rater agreement indices (r_{WG}), which show evidence of consensus amongst the respondents (James et al., 1984).

Determining exposure of the workforce to CPFs

It has been argued in Manu et al. (2012) that in the context of CPFs, due to their remoteness in the process of accident causation, workforce exposure to their potential to influence accident occurrence can realistically be assessed at a generic project level, such as in the form of the duration within which a CPF applies to a project, or by broadly assessing exposure in terms of whether or not a CPF applies to a project. However, given the difficulty in assessing exposure regarding duration for some CPFs (e.g. level of construction, and subcontracting), assessing exposure

Table 4.1 Potential of CPFs to influence accident occurrence

Construction project feature	Mean potential to influence accident occurrence*	Standard deviation	Minimum	Maximum	r_{WG}**
Demolition	3.1739	.95367	0.00	4.00	0.55
Underground construction	2.8368	.89677	0.00	4.00	0.60
Tight project duration	2.8361	.70531	0.00	4.00	0.75
High-level construction	2.7554	.89319	0.00	4.00	0.60
Multi-layer subcontracting	2.6998	.78400	0.00	4.00	0.69
Complex design	2.6141	.84802	0.00	4.00	0.64
Restricted site	2.6089	.80872	0.00	4.00	0.67
Restricted site locality	2.5703	.76306	0.00	4.00	0.71
Refurbishment	2.5169	.92349	0.00	4.00	0.57
Traditional on-site construction	2.2174	.65830	0.00	4.00	0.78
New work	1.9858	.75112	0.00	4.00	0.72
Management contracting	1.9499	.76143	0.00	4.00	0.71
Design-and-build procurement	1.8260	.77698	0.00	4.00	0.70
Traditional method of procurement	1.8058	.81008	0.00	4.00	0.67
Unrestricted site locality	1.7955	.74548	0.00	4.00	0.72
Unrestricted site	1.7949	.78860	0.00	4.00	0.69
Partnering procurement	1.7709	.76016	0.00	4.00	0.71
Low-level construction	1.7111	.73799	0.00	4.00	0.73
Adequate project duration	1.6558	.72922	0.00	4.00	0.60
Single-layer subcontracting	1.6252	.72704	0.00	4.00	0.74
Simple design	1.5475	.73703	0.00	3.00	0.73
Pre-assembly construction	1.5146	.77634	0.00	4.00	0.70

* Mean ratings are based on a 5-point scale (0 = none, 1= low, 2= moderate, 3 = high, 4 = very high).
**r_{WG}= *single-* item inter-rater agreement index. r_{WG} indices are based on a uniform null distribution. Based on 10,000 simulation runs. r_{WG} values of 0.08, 0.10, and 0.14 are the 90%, 95% and 99% confidence interval estimates, respectively, for group size of 184 and five response options (i.e. a 5-point scale). Hence, r_{WG} values > 0.14 are evidence of significant agreement at $p < 0.01$.

in terms of whether or not a CPF applies to a project is a more practical option. This means that if a CPF applies to a project, the workforce will be exposed to its potential to influence accident occurrence, and where a particular CPF does not apply to a project, the workforce will not be exposed to its potential to influence accident occurrence. In line with semi-quantitative risk evaluation, where qualitative information is assigned numeric scales, assessing exposure in this manner can be expressed as a binary situation, where a no-exposure condition is assigned a 0, and a condition where the workforce is exposed is assigned a 1. It is logical to assign a 0 to a no-exposure situation, because without exposure there can be no risk (Duffus and Worth, 2001; HSE, 2001), and, as such, any degree of potential to influence accident occurrence combined (i.e. multiplied) with a no-exposure condition will yield a risk of zero.

Definition of risk categories in the risk combination matrix

The degree of potential of CPFs to influence accident occurrence has been assessed using a 5-point scale (where 0 = none, 1 = low, 2 = moderate, 3 = high, and 4 = very high). Having determined a semi-quantitative scale for expressing exposure (i.e. 0 = workforce not exposed – where a CPF does not apply to a project – and 1 = workforce exposed – where a CPF applies to a project), the risk expression (i.e. *Eqn. 1*) was then applied using a risk combination matrix, as presented in Table 4.2 below.

From the matrix shown in Table 4.2, it is evident that the H&S risk associated with a CPF can range from a score of 0 (representing the lowest level of risk) to 4 (representing the highest level of risk). These numeric levels of risk need to be assigned qualitative risk categories to enable interpretation of the risk matrix. In this regard, the risk categorisation proposed by the BSI (2008) was very instructive. The BSI (2008) proposed a five-band risk categorisation and acceptability rating, as follows.

- Very low: Risk is considered acceptable. No further action is necessary other than to ensure that the controls are maintained.
- Low: No additional controls are required unless they can be implemented at very low cost (in terms of time, money, and effort). Actions to reduce these risks are assigned low priority.
- Medium: Consideration should be given to whether the risk can be lowered, but the cost of additional risk-reduction measures should be taken into account.
- High: Substantial efforts should be made to reduce the risk. Risk reduction measures should be implemented urgently.
- Very high: Risk is unacceptable. Substantial improvements in risk controls are necessary so that the risk is reduced to an acceptable level.

For assessment of the H&S risk associated with CPFs, this five-band categorisation was adapted. Applying this categorisation to Table 4.2, as no exposure

Table 4.2 Risk combination matrix

		Exposure	
		0	1
Potential to influence	0	0	0
accident occurrence	1	0	1
	2	0	2
	3	0	3
	4	0	4

Table 4.3 Risk combination matrix, with defined risk categories

		Exposure	
		0	1
Potential to influence	0	0 (No risk)	0 (No risk)
accident	1	0 (No risk)	1 (Low risk)
occurrence	2	0 (No risk)	2 (Medium risk)
	3	0 (No risk)	3 (High risk)
	4	0 (No risk)	4 (Very high risk)

results in no risk, and, similarly, an absence of potential to cause harm also results in no risk (Duffus and Worth, 2001; HSE, 2001), the "0" risk values in Table 4.3 were assigned the *"No risk"* category. Risk score "4", being the highest score, was assigned the *"Very high risk"* category, and risk score "2", being the mid-score, was assigned the *"Medium risk"* category. Risk score "1", being the next risk score below risk score "2", was assigned the *"Low risk"* category, and risk score "3", being the next risk score above risk score "2", was assigned the *"High risk"* category. Assigning the qualitative risk categories to the numeric scores results in an interpretable risk combination matrix, as presented in Table 4.3. Following the systematic definition of the risk categories, the H&S risk associated with CPFs was assessed by combining the mean scores of the degree of the potential of the CPFs to influence accident occurrence (in Table 4.1) and the binary scores for exposure. The risk assessment that resulted from this is discussed in the following section.

Results and discussion

The outcome of the risk combination is given in Table 4.4. From the assessment, where a CPF applies to a project, demolition has the highest degree of risk (risk score 3.17), and pre-assembly construction has the lowest degree of risk (risk score 1.51). To assist with better interpretation of the numeric risk scores, they are approximated to the nearest risk score, to give an overall risk category. From Table 4.4, the degree of H&S risk associated with a CPF where it applies to a project can be considered as either being *high-risk* or *medium-risk*. Amongst the 22 CPFs, 9 are *high-risk* features, implying that they are associated with a high likelihood of accident occurrence. The remaining 13 CPFs are *medium-risk* features, implying that they are associated with the *medium* likelihood of accident occurrence. The overall risk assessment is consistent with some of the few qualitative comparative risk evaluations in the literature, such as new work being considered as having less risk than refurbishment (Anumba et al., 2006). However, contrary to pre-assembly construction being considered as having less degree of H&S risk than traditional methods of construction (see McKay et al., 2002), the assessment presented in Table 4.5 indicates that pre-assembly and traditional construction are similarly medium-risk CPFs. In terms of

Table 4.4 Risk combination matrix for evaluation of the H&S risk associated with CPFs

Construction project feature	Potential to influence accident occurrence	Exposure		H&S risk associated with CPF where CPF applies to a project	
	Mean score	0 (CPF does not apply to a project)	1 (CPF applies to a project)	High risk (3)	Medium risk (2)
Demolition	3.17	0	3.17	✓	
Underground construction	2.84	0	2.84	✓	
Tight project duration	2.84	0	2.84	✓	
High-level construction	2.76	0	2.76	✓	
Multi-layer subcontracting	2.7	0	2.7	✓	
Complex design (i.e. design with intricate aesthetic qualities)	2.61	0	2.61	✓	
Restricted site (i.e. where footprint of facility covers most of the site area)	2.61	0	2.61	✓	
Restricted site locality, e.g. city-centre location	2.57	0	2.57	✓	
Refurbishment	2.52	0	2.52	✓	
Traditional on-site construction	2.22	0	2.22		✓
New work	1.99	0	1.99		✓
Management contracting	1.95	0	1.95		✓
Design-and-build procurement	1.83	0	1.83		✓
Traditional method of procurement	1.81	0	1.81		✓
Unrestricted site locality	1.8	0	1.8		✓
Unrestricted site	1.79	0	1.79		✓
Partnering procurement	1.77	0	1.77		✓
Low-level construction	1.71	0	1.71		✓
Adequate project duration	1.66	0	1.66		✓
Single-layer subcontracting	1.63	0	1.63		✓
Simple design	1.55	0	1.55		✓
Pre-assembly construction	1.51	0	1.55		✓

Table 4.5 Tool-verification feedback

PART A: Respondents' information

Respondent's ID	Role	Years of experience in construction
1	H&S manager	9
2	Construction H&S consultant	15
3	H&S manager	22
4	Business improvement manager	40
5	Construction manager	13
6	H&S manager	25
7	H&S manager	9
8	H&S manager	9
9	Construction H&S consultant	30
10	H&S manager	30
11	H&S manager	30
12	Construction manager	30
13	H&S manager	30

Part B: Similarity of H&S risk

	Response				
	No response	Not similar	Slightly similar	Fairly similar	Similar	Very similar
1. How similar is the risk profile to the H&S challenges and risk experienced on the project?	15.38% (2)	0% (0)	15.38% (2)	30.76% (4)	23.10% (3)	15.38% (2)

Note: The figure in brackets represents the number of respondents.

(Continued)

Table 4.5 Tool-verification feedback (Continued)

Part C: *Usefulness of tool to pre-construction H&S planning*	*Response*						
	No response	*Not useful*	*Slightly useful*	*Fairly useful*	*Useful*	*Very useful*
2. At the concept/design stage of a new development, how useful could risk information such as that provided by this tool be regarding informing/ influencing decisions that determine CPFs?	18.18% (2)	0% (0)	9.09% (1)	36.36% (4)	18.18% (2)	18.18% (2)
3. At the concept/design stage of a new development, how useful could risk information such as that provided by this tool be concerning informing/ influencing the planning of design or project management solutions to control H&S risk posed by CPFs?	0% (0)	0% (0)	9.09% (1)	45.45% (5)	27.27% (3)	18.18% (2)

Note: Total number of applicable respondents = 11. The figure in brackets represents the number of respondents. Applicable respondents are those whose organisations' operations are likely to involve early (e.g. concept/design-stage) H&S planning, e.g. housing development.

acceptability of risk, referring to guidelines by the BSI (2008), none of the CPFs is associated with acceptable risk, indicating that whichever CPFs apply to a project, measures need to be taken to mitigate their associated risk, where the extent of the measures should depend on the degree of risk. While for medium-risk CPFs consideration should be given to whether the risk can be lowered, for high-risk CPFs substantial measures are required to be implemented.

Development and verification of a simple pre-construction phase project H&S risk-profiling tool

Tool development

The features of projects can vary from project to project, and, as such, the features of a project, say Project "A" (e.g. refurbishment, high-level construction, tight duration, etc.), can differ from those of another project, say Project "B" (e.g. new build, low-level construction, complex design, etc.). It is thus desirable to have a simple tool that will enable project participants (at pre-construction stage) to select the features that apply or would apply to a project and to view an overall H&S risk profile for the project (based on the selected/applicable features). Given this, a simple computer-based tool using Microsoft Excel was developed to represent the risk information given in Table 4.5. Microsoft Excel, amongst its several functions for executing tasks, has a function called the "IF function" (Bluttman and Aitken, 2010). The IF function returns a particular value if a condition is true, and another value if that condition is false (Bluttman and Aitken, 2010). This function can be very useful where, among several possible scenarios, a user wishes to display only the specific information relevant to a particular scenario. As different CPFs may apply to any single project, this function in Microsoft Excel can be used to create a series of instructions which retrieve and display only the H&S risk profile related to a selection of CPFs which apply to a project. This function in Microsoft Excel was thus used extensively in creating a simple pre-construction phase project risk-profiling tool. In addition to displaying the H&S risk profile of a project, the tool included suggestions for mitigating the risk associated with the CPFs. The recommendations are based on the hierarchy of risk control (BSI, 2008; Hughes and Ferrett, 2008) and the effect of two factors that have been shown to affect the potential of a CPF to influence accident occurrence: (1) the potential of proximate accident cause(s) that are related to the CPF to influence accident occurrence; and (2) the degree of prevalence of the proximate accident cause(s) within a CPF (Manu et al., 2014). For example, high-level construction is related to the proximate accident cause "working at height" (see Chapter 3). According to Manu et al. (2014), the potential for high-level construction to influence accident occurrence is therefore related to the potential for "working at height" to influence accident occurrence (i.e. the potential for working at height to cause harm), and the degree to which "working at height" is prevalent within high-level construction. To mitigate the level of H&S risk associated with high-level construction, there is therefore a need to either (1) mitigate the potential of

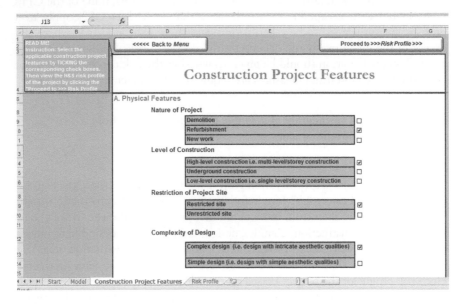

Figure 4.1 A print screen depiction of "CRiMT"

"working at height" to cause harm, by making it safe for operatives to work at height (e.g. using fall protection) or (2) mitigate the prevalence of "working at height" (i.e. reducing the need for operatives to work at height, by designing out work at height), or both. The Microsoft Excel tool has been nicknamed "CRiMT". Figure 4.1 provides a print screen depiction of "CRiMT".

Tool verification

To enable verification of the risk information in Table 4.4 and the industrial relevance of the tool to pre-construction H&S planning, the tool was tested by experienced construction professionals in construction management-related roles. The professionals were asked to trial the tool by selecting the features of a recently completed project they have worked on, and then proceeding to view the H&S risk profile displayed by the tool. They were then asked to respond to three main questions. The first question was about the similarity between the displayed H&S risk profile and the actual H&S risk experienced on the project. The remaining two questions which explored the relevance of the tool to pre-construction H&S planning were only applicable to those respondents whose organisations' operations involve early (e.g. concept/design-stage) H&S planning, e.g. housing development, design-and-build, etc. The demographic information of the professionals (n = 13) is provided in Part A of Table 4.5 below.

The minimum and the mean number of years of experience of the respondents in construction are nine years and 22.46 years, respectively. Collectively, this indicates a respectable amount of experience in the management of construction and H&S (see Hallowell and Gambetese, 2010), and therefore the respondents' feedback can be deemed credible. In terms of how the risk profile given by the toolkit compares with the H&S challenges and risk experienced on the respondents' projects, as shown by Part B of Table 4.5, the responses range from "slightly similar" to "very similar", with a majority of the respondents (i.e. 9, representing 69.23%) indicating at least fair similarity. This means that the risk information given by the tool reasonably reflects the actual H&S risk situations induced by the CPFs on the respondents' projects. The responses (given in Part C of Table 4.5) indicate that for 11 of the 13 respondents, the nature of their organisations' operations involves pre-construction decision-making, which, among other things, determines CPFs, and also planning design or project management solutions to control H&S risk posed by CPFs. These respondents include two H&S consultants, whose role involves advising on such matters. As can be seen from Table 4.5, a majority of the respondents (i.e. 8 out of the 11, representing 72.73%) indicated that the information given by the toolkit is at least fairly useful for informing/influencing decisions that determine CPFs. Again, a majority of the respondents (i.e. 10 out of the 11, representing 90.91%) indicated that the information given by the tool is at least fairly useful for informing/influencing the planning of design or project management solutions to control H&S risk posed by CPFs. Overall, these results indicate that the risk information given by the tool is of practical relevance to managing the H&S risk posed by CPFs during the pre-construction stage of project delivery.

Implications

In the light of the overall risk assessment, *high-risk* CPFs (e.g. tight project schedules and intricate designs) should certainly not be taken lightly. For instance, clients ought to allow sufficient time and resources for projects, and designers ought to produce designs that will be safe for contractors to build and maintain. In the UK this is required under the CDM regulations. Adhering to such requirements should not be taken as being trivial or as a mere checklist exercise, as there is a great likelihood of accidents occurring on projects where these features and similar *high-risk* features apply. As CPFs emanate from the pre-construction stage of project procurement, through decisions by the client, the design team, and the project management team, these project participants have an enormous opportunity to mitigate the H&S risk associated with projects by their characteristics. In making decisions that determine CPFs, the decision-makers could select *medium-risk* CPFs over *high-risk* CPFs, as there is relatively less likelihood of accident occurrence associated with *medium-risk* CPFs. As the selection of a *high-risk* CPF may be inevitable in some situations, due to possible project constraints, such as client requirements (Suraji et al., 2001), pre-construction project participants will have to implement risk-mitigation measures. In doing so, greater priority ought to be given to *high-risk* CPFs, and, as such, substantial efforts (i.e. resources) will have to be allocated towards mitigating the risk associated with *high-risk* CPFs. With medium-risk CPFs, as there is still some likelihood of accident occurrence where such CPFs apply, they cannot be completely disregarded. As recommended by the BSI (2008), further consideration ought to be given to whether the risk can be reduced, while taking into account the cost of any measures.

In all, the risk assessment accords each CPF its degree of H&S risk, which allows for comparison between all the CPFs. The risk assessment offers important insights, which should assist project participants in ascertaining the H&S risk profile of projects at the pre-construction stage. The risk assessment could also inform pre-construction decision-making regarding CPFs, as well as prioritisation of risk-control measures. In this light, CRiMT could be a useful resource, particularly to project participants who may have limited experience in managing the H&S risk posed by certain features of construction projects.

Summary

Ascertaining the H&S risk profile of a construction project can be difficult, particularly at the pre-construction stage of a project, due to the often limited information available at that stage. It may also be difficult for some project participants if they have limited experience in delivering a particular kind of project. However, based on some attributes of projects which emanate from pre-construction decisions and also influence accident occurrence, it can be possible to gauge the H&S risk profile of a project at the pre-construction stage. In this chapter, the degree of H&S risk associated with construction project features (CPFs) (e.g. the nature of the project, the method of construction used, the project duration, the level of construction, whether subcontracting is involved, the design complexity, the type of site restriction, and the procurement system used) have been evaluated using a semi-quantitative risk-evaluation technique and an adapted mathematical H&S risk expression. Quantitative inputs for the expression were partly provided by the results of a questionnaire survey. The risk evaluation shows that the CPFs are associated with *high risk* or *medium risk*, implying that they are associated with a high likelihood of accident occurrence or a medium likelihood of accident occurrence, respectively. While with medium-risk CPFs some risk-control measures will suffice in mitigating risk, with high-risk CPFs substantial measures are required. These findings have implications for pre-construction decision-making regarding CPFs, and also for prioritisation of risk-control measures. CRiMT, a simple H&S risk-assessment tool developed from the evaluation, could be a useful resource to pre-construction project participants in ascertaining the H&S risk profile of a construction project based on its applicable/relevant features, to guide the introduction of appropriate H&S risk control measures. Construction accident causation, being a complex phenomenon, can, however, involve inter-causal relationships (see Chapter 3) between causal factors, which could affect the H&S risk associated with a CPF. One of the limitations of CRiMT is that the potential effect of inter-causal relationships on the H&S risk associated with a CPF is not taken into account. Also, beyond the CPFs that were investigated in this chapter, there may be other features of projects (determined by pre-construction decisions) that also have an accident causal role. The CPFs included in *CRiMT* are therefore not exhaustive. *CRiMT* should therefore not be viewed as a static tool, but rather as a dynamic tool, for which subsequent versions are incorporating new insights can be developed.

References

Ankrah, N.A. (2007). *An investigation into the impact of culture on construction project performance*. Ph.D. thesis. The University of Wolverhampton.

Anumba, C., Egbu, C. and Kashyap, M. (2006). *Avoiding structural collapses in refurbishment: A decision support system*. Suffolk: HSE Books.

Aven, T. (2008). A semi-quantitative approach to risk analysis, as an alternative to QRAs. *Reliability Engineering & System Safety*, 93(3): 768–775.

Bluttman, K. and Aitken, P.G. (2010). *Excel formulas and functions for dummies*. 2nd edn. Hoboken, NJ: Wiley.

BOMEL Limited, Glasgow Caledonian University and The Institute for Employment Research (2006). *An analysis of the significant causes of fatal and major injuries in construction in Scotland (Factors influencing Scottish construction accidents – FISCA)*. Research Report 443. Suffolk: HSE Books.

Brabazon, P., Tipping, A., and Jones, J. (2000). *Construction health and safety for the new Millennium*. Suffolk: HSE Books.

Brace, C., Gibb, A., Pendlebury, M. and Bust, P. (2009). *Phase 2 Report: Health and safety in the construction industry: Underlying causes of construction fatal accidents –External research*. Norwich: Her Majesty's Stationery Office.

British Standard Institute (2008). *Guide to achieving effective occupational health and safety performance*. BS 18004:2008. London: British Standard Institute.

BSI *see* British Standard Institute.

Canadian Centre for Occupational Health and Safety (2008). *Risk versus hazards*. Hamilton, ON: Canadian Centre for Occupational Health and Safety.

Chicken, J.C. and Posner, T. (1998). *The philosophy of risk*. London: Thomas Telford.

Chua, D.K.H. and Goh, Y.M. (2005). Poisson model of construction incident occurrence. *Journal of Construction Engineering and Management*, 131(6): 715–722.

Cooke, T. and Lingard, H. (2011). A retrospective analysis of work-related deaths in the Australian construction industry. In: Egbu, C. and Lou, E.C.W. (eds), *Proceedings of the 27th Annual ARCOM Conference, 5-7 September 2011*. Bristol, UK: Association of Researchers in Construction Management. pp. 279–288.

Department of Occupational Safety and Health (2015). Occupational accidents statistics by sector until December 2015. Department of Occupational Safety and Health. Available online at http://www.dosh.gov.my/index.php/en/archive-statistics/2015/1713-occupational-accidents-statistics-by-sector-until-december-2015 (Accessed 12 July 2016).

Duffus, J. and Worth, H. (2001). *The science of chemical safety essential toxicology: An educational resource*. Research Triangle Park, NC: International Union of Pure and Applied Chemistry.

Egbu, C.O. (1999). Skills, knowledge and competencies for managing construction refurbishment works. *Construction Management and Economics*, 17(1): 29–43.

Hallowell, M.R. and Gambatese, J.A. (2010). Qualitative research: Application of the Delphi method to CEM research. *Journal of Construction Engineering and Management*, 136(1): 99–107.

Hide, S. (2003). *Exploring accident causation in the construction industry*. PhD thesis. Department of Building and Civil Engineering, Loughborough University.

Hide, S., Atkinson, S., Pavitt, T., Haslam, R., Gibb, A., Gyi, D, Duff, R. and Suraji, A. (2003). *Causal factors in construction accidents*. Suffolk: HSE Books.

Horbury, C. and Hope, C. (1999). *The impact of procurement and contracting practices on health and safety – A review of literature*. Report RAS/99/02. Buxton: HSL.

HSE (2000). *Management of Health and Safety at Work Regulations 1999. Approved Code of Practice & Guidance*. 2nd edn. Suffolk: HSE Books.

HSE (2001). *Reducing risks: Protecting people – HSE's decision making process*. Norwich: Her Majesty's Stationery Office.

HSE (2006). *Five steps to risk assessment*. Suffolk: HSE Books.

HSE (2009). *Construction intelligence report: Analysis of construction injury and ill-health intelligence*. Health and Safety Executive. Available online at http://s3.amazonaws.com/zanran_storage/www.hse.gov.uk/ContentPages/25543732.pdf (Accessed 1 January 2011).

HSE (2015a). *Historical picture – HISTINJ – Reported injuries in Great Britain by main industry and severity of injury, 1974 to latest year*. HSE. Available online at www.hse.gov.uk/Statistics/tables/index.htm (Accessed 12 July 2016).

HSE (2015b). *Managing health and safety in construction – CDM 2015 Guidance L153*. HSE.

Hughes, P. and Ferrett, E. (2008). *Introduction to health and safety in construction*. 3rd edn. Oxford: Elsevier.

ILO (2005). *Global estimates of fatal work related diseases and occupational accidents, World Bank Regions*. Geneva: ILO.

James, L.R., Demaree, R.G. and Wolf, G. (1984). Estimating within-group interrater reliability with and without response bias. *Journal of Applied Psychology*, 69(1): 85–98.

Jannadi, O.A. and Almishari, S. (2003). Risk assessment in construction. *Journal of Construction Engineering and Management*, 129(5): 492–500.

Kariuki, S. and Löwe, K. (2007). Integrating human factors into process hazard analysis. *Reliability Engineering & System Safety*, 92(12): 1764–1773.

Knab, L.I. (1978). Numerical aid to reduce construction injury losses. *Journal of the Construction Division*, 104(4): 437–445.

Manu, P., Ankrah, N., Proverbs, D. and Suresh, S. (2012). Investigating the multi-causal and complex nature of the accident causal influence of construction project features. *Accident Analysis and Prevention*, 48: 126–133.

Manu, P., Ankrah, N., Proverbs, D. and Suresh, S. (2013). Mitigating the health and safety influence of subcontracting in construction: The approach of main contractors. *International Journal of Project Management*, 31(7): 1017–1026.

Manu, P., Ankrah, N., Proverbs, D. and Suresh, S. (2014). The health and safety impact of construction project features. *Engineering, Construction and Architectural Management*, 21(1): 65–93.

Mayhew, C. and Quinlan, M. (1997). Subcontracting and occupational health and safety in the residential building industry. *Industrial Relations Journal*, 28(3): 192–205.

McKay, L.J., Gibb, A.G.F., Haslam, R. and Pendlebury, M. (2002). Implications for the effect of standardization and pre-assembly on health, safety and accident causality-preliminary results, in *18th Annual ARCOM Conference*, 2-4 September 2002, University of Northumbria. Association of Researchers in Construction Management.

Ministry of Manpower (2015). *A healthy workforce in a safe workplace – Annual Report 2015*. Ministry of Manpower. Available online at www.mom. gov.sg/ebook/oshd-ar2015/pdf/OSHD_AR2016_LoRes.pdf (Accessed 12 July 2016).

Perttula, P., Merjama, J., Kiurula, M. and Laitinen, H. (2003). Accidents in materials handling at construction sites. *Construction Management and Economics*, 21(7): 729–736.

Pinto, A., Nunes, I.L. and Ribeiro, R.A. (2011). Occupational risk assessment in construction industry – Overview and reflection. *Safety Science*, 49(5): 616–624.

Risk & Policy Analyst Ltd (1999). *Risk ranking for small and medium enterprises*. Suffolk: HSE Books.

Sachs, T. and Tiong, R.L.K. (2009). Quantifying qualitative information on risks: Development of the QQIR method. *Journal of Construction Engineering and Management*, 135(1): 56–71.

Sacks, R., Rozenfeld, O. and Rosenfeld, Y. (2009). Spatial and temporal exposure to safety hazards in construction. *Journal of Construction Engineering and Management*, 135(8): 726–736.

Smith, N.J., Merna, T. and Jobling, P. (2006). *Managing risk in construction projects*. 2nd edn. Oxford: Blackwell.

Strategic Forum for Construction (2002). *Accelerating change: A report by the Strategic Forum for Construction, chaired by Sir John Egan*. London: Rethinking Construction.

Suraji, A. (2001). *Development and validation of a theory of construction accident causation*. PhD Thesis. University of Manchester Institute of Science and Technology (UMIST).

WHO and FAO (2009). *Risk characterization of microbiological hazards in food: Microbiological risk assessment series 17*. WHO and FAO.

Wright, M., Bendig, M., Pavitt, T. and Gibb, A. (2003). *The case for CDM: Better safer design – a pilot study*. Suffolk: HSE Books.

5 The impact of an ageing workforce on the construction industry in Australia

Alpana Sivam, Tony Trasente, Sadasivam Karuppannan and Nicholas Chileshe

Experience without theory is blind, but theory without experience is mere intellectual play–

(Immanuel Kant)

Background

The increase in life expectancy over recent decades has led to substantial population ageing occurring throughout the world (Garin et al., 2014). The proportion of the population aged 60 and above is increasing (WHO, 2007) and growing at a rate of 3.26 per cent per year (United Nations, 2015). Globally, the number of persons aged 60 and above is expected to more than double by 2050 and more than triple by 2100, increasing from 901 million in 2015 to 2.1 billion in 2050 and 3.2 billion in 2100 (United Nations, Department of Economic and Social Affairs, Population Division, 2015). These demographic changes are posing social, economic and personal challenges for society, families, and individuals (Sivam, 2011). The change of the age structure of the population will have an immense impact on the composition of the workforce globally. The extent of the impact will depend on the context and industry and will not necessarily be the same across the world.

Currently, Australia is experiencing rapid demographic change which is contributing to the ageing workforce in all sectors of industry. The population of Australia is ageing and it is projected that the number of people aged 65 and above will have increased from 13 per cent in 2010 to 23 per cent by June 2050 (Australian Physiotherapy Association, 2016). It has also been noticed by some researchers that the Australian workforce is retiring earlier than workers in comparable countries. Watson (2012) states that "...in Australia, only 49 per cent of people between the age of 55 and 64 are in the workforce, compared to 59 per cent in the US". However, ageing in construction is defined based on diminished physical capacity rather than chronological age (Marchant, 2013).

The construction industry is a significant driver of both economic activity as well as the social and physical well-being of Australia through the construction of housing and commercial buildings, social and physical infrastructure and transportation (Construction Training Council, 2014). In 2014–15 the construction industry contributed 7.8 per cent to Australia's gross domestic product (GDP), up from 6.5 per cent a decade previously (Australian Industry Group, 2015). It is Australia's third largest industry, behind only mining (GDP 8.8 per cent) and finance (8.7 per cent). It comprises over 330,000 businesses nationwide and directly employs over one million people (around 9 per cent of the total workforce). It produces buildings and infrastructure that are essential to the operation of all industries and add to the wealth and capital stock of the nation, and underpins the productivity improvements that are necessary to support future prosperity and well-being.

The nature of the construction industry is characterised by three critical sectors, namely housing, commercial and civil engineering operations. There are over 100 different types of occupations in the industry. These include management, professional, para-professional, administrative, trade, technical and semi-skilled workers. The tradesmen and technicians account for the largest number of workers, namely approximately 49 per cent of the total workforce in the construction sector. Compared to other industries, the construction industry heavily relies on the recruitment of apprentices and has traditionally been an active trainer of young workers. By 2034/35 almost 20 per cent of the Australian population (6.2 million) is projected to be aged 65 or over. The construction sector also has a high percentage of workers aged 55 and above. There is a growing concern in Australia about the long-term shortage of skilled construction tradespeople because not enough apprentices are being trained to replace the retiring workers in order to meet the increasing demand for new labour (Heaton, 2015). Greater and more flexible involvement of older workers and changes in policy may help overcome part of the problem (Watson, 2012). The construction industry in Australia is currently undergoing revolutionary changes.

Research has shown that the construction industry in Australia will face some challenges in adjusting to the decline in the growth of the labour force due to the ageing of the labour force. The skills' shortage will affect the overall performance and productivity of construction industry throughout Australia (Watson, 2012). The reasons for skills' shortages and gaps are often complex with the specific causes varying with the industry and occupation (Watson, 2012: 45). There is a diverse range of factors contributing to current and future skills' shortages. These include an ageing workforce and forthcoming retirement rates; changing skills required for various occupations; differences in demand for and supply of skilled workers as a result of employment arrangements; poor educational qualifications translating into smaller and lower numbers of successful job applicants; inadequate apprenticeship rates and difficulty in attracting and retaining employees (Watson, 2012).

Globally the number of older workers will grow, while in developed countries the share of younger workers will decline, leading to a shortage of workers. Even though developed economies will attract foreign workers as a result of globalisation, these will not be the same as local workers. Local labour brings in-depth construction techniques that the construction industry can use to its advantage and thus avoid training overseas workers to meet the specific needs of local industry. Therefore, this chapter aims to identify issues of skills' shortage due to the ageing of the workforce and evolve a viable road map to address issues facing the construction industry.

The chapter is divided into three parts. The first section presents an overview of the demographics of ageing and the ageing of the workforce, and reviews the skills' shortage in the construction industry. The next section presents the research framework, followed by results and findings.. Following that, the road-map that can be used as a vehicle to mitigate the imbalance between the supply of and demand for skills in the construction industry due to the ageing of the workforce is presented. Finally, the conclusion is presented to address the short-age of skilled labour in the construction industry, as well as limitations and opportunities for future research.

Conceptual foundation

There is a basic relationship between wages and productivity. According to van Dalen et al. (2010), whenever an ageing workforce is involved, the basic concern of the employer is the wage-productivity gap. If the wages of aged workers exceed their productivity, aged workers present a loss for the company. It is based on the human capital theory that looks into the relationship between productivity and age (Becker, 1962). Productivity is directly related to human capital. Depreciation of human capital is very much like physical capital. To improve human capital, either the firm or workers have to invest in human capital (van Dalen et al., 2010). Therefore it is relevant to understand the conceptual background of demographic ageing, the ageing workforce and the workforce profile in the construction industry.

Demographic ageing and ageing workforce

Ageing is a fact of life and questions about ageing are significant in order to understand this aspect of life (Marchant, 2013). It is a biological phenomenon, not a social construct, but it has a great impact on human life through, firstly, the process of physiological decline, thereby increasing the incidence of disability; secondly, a decline in employment opportunities; and thirdly, a change in social and economic circumstances (Arber et al., 2003).

The ageing population is one of the most prominent global demographic trends in the history of our society. The world's population is ageing at unprec-edented levels (O'Hehir, 2014). According to Turnbull (1990), the term 'the aged' officially refers to those people who qualify for the state Pension, at 65 years for men and 60 years for women but in general, 'the aged' are those people who suffer from connotations of dependency, senescence and physical and psycho-logical deprivation. Policy makers and demographers categorise 'the aged' as those aged 55 and above due to the increasing proportion of the population entering retirement age and the lowering of the retiring age, in some cases up to

55 years (Turnbull, 1990). Though with the passage of time there may be an increase of physical, mental and functional problems in humans, chronologically old age is not a phase of decline and loss as it provides the opportunity for positive change and productive functioning when approached correctly (Balcombe & Sinclair, 2001). Thus longevity is the result of socioeconomic and technological developments and can provide opportunities for growth. Societies where the elderly have a good quality of life will be healthy communities for any age group. Unlike financial crises or natural disasters, ageing trends and their impact can be predictable and hence policies for an ageing society should be developed. Policies for such societies are thus not only about targeting present needs and opportunities, but also about anticipating future population structures and pathways for a smooth transition (OECD, 2015).

Globally, the proportion of the population aged 60 years or over is predicted to double from 11 per cent in 2006 to 22 per cent by 2050 (O'Hehir, 2014; Plouffe & Kalache, 2010). In 2015, one out of every eight persons worldwide was aged 60 or over, while by 2030 the population aged over 60 is projected to account for one in six people in the world (United Nations, 2015). By the middle of the twenty-first century, one in every five people will be aged 60 years or over (UN, 2015: 3). The significant escalation in the world's aged population will affect the construction industry owing to an ageing workforce. In western countries such as Australia, this trend will accelerate over the next two decades as the Baby Boomers continue to move into the over 65 age group. All industrialised countries have an ageing workforce (Armstrong-Stassen & Templer, 2005). Along with many western nations, Australia's population is ageing, and it is predicted that within the next 50 years there will be double the number of pensioners (Frommert et al., 2009).

Since 1990 the number of workers aged over 40 has increased by 2 per cent (Cook et al., 2009). A recent study conducted by the Chartered Institute of Building (CIOB) (2014) shows that the 100+ age group is growing in many developed countries such as the UK and the USA. Globally the number of older workers will increase but in developed countries the share of young workers will decline, leading to a shortage of employees. Owing to globalisation, developed countries will attract and train overseas workers, but these will differ from the local workers because of their knowledge of construction techniques that the construction industry can use to its advantage, thereby minimising the risk of mistakes being made.

Even though the ageing of a workforce is a significant issue, the definition of older workers is fuzzy (Marchant, 2013). In management literature, older workers are defined as workers aged 40 or 45 (Marchant, 2013) whereas the Australian Bureau of Statistics identifies mature age workers as those who are 45 and older (ABS, 2005). Other practical definitions include the age at which workers' compensation is no longer available, which is usually 65. To retain the ageing workforce, the Australian government has proposed to raise the qualifying age for Age Pensions from 65 to 67 for both men and women between 2017 and 2023 (ABS, 2011).

Ageing will not only impose challenges for the construction sector but also for health, social and housing policies. It also has financial implications, as well as implications for employment, public expenditure industry output and social relationships. However, the focus of this chapter is on the construction sector.

Workforce profile in construction industry

The construction industry plays a significant role in economic growth and employment output and is well known for being a manual and physically demanding industry. The construction industry is dominated by male workers and self-employed tradespeople. It is vital to comprehend retention and minimise the risks faced by old male workers (McNair & Flynn, 2006). From an epidemiological perspective, work-related injuries of older workers are serious. Research conducted in the US does not indicate a greater number of injuries but rather injuries of a more severe nature (Schwatka et al., 2012).

Construction is the third largest industry in Australia, employing more than one million workers (9 per cent) nationally next to health care and social assistance (13 per cent) and retail trade (11 per cent) (Department of Employment, 2016). The construction sector operates in both the private and public sectors, mainly in three categories: engineering construction (major infrastructure, mining and heavy industrial resource-based projects); non-residential building (including offices, shops, hotels, industrial premises, hospitals, and entertainment facilities); and housing (houses, flats, home units, and townhouses).

The workforce participation rate of the population aged 55 and over in Australia has increased from 25 per cent to 34 per cent over the past 30 years, and most of the increase occurred in the last decade (ABS, 2010). This will have an important effect on the industry as essential skills will be lost when employees retire. This is exacerbated by the fact that the number of new recruits is declining, and there will be fewer people to replace those retiring. Worldwide, skilled trades are the hardest positions to fill (Manpower, 2010). Extending the retirement age may not offer the solution the construction industry needs. Many construction jobs are physically demanding, and so it is not always possible to safely extend the retirement age for all trades related to the construction sector. In 2012, just three construction trades were in short supply compared to 12 in 2008 (Department of Employment, Education and Workforce Relations, 2012). In the longer term, with a significant number of major infrastructure and resources projects on the agenda in several Australian states and territories, a substantial and sustained demand for trade skills will be expected (Marchant, 2013).

Research on ageing in the construction industry is scarce. The sector needs to become more age responsive because of a talent shortage. According to Marchant (2013), older workers have basic knowledge sets that could be harnessed for mentoring and training but to explore how to make this happen needs more research.

Research framework

The previous sections have indicated the context of a shortage of construction labour due to an ageing workforce and the challenge facing the construction industry. This section describes the methodology undertaken for this research, including the data collection methods. A content analysis approach has been adopted. Content analysis was used because it allows the examining of trends and pattern in documents (Stemler, 2001; Mayring, 2000). Researchers have used content analysis either as a qualitative or quantitative method (Berelson, 1952 quoted in Mayring, 2000). In this research, content analysis is used as a qualitative research technique to analyse qualitative and quantitative data. Since the main focus is on how the workforce profile in construction is changing and its subsequent impact on the construction industry, this approach was deemed most suitable. Content analysis is a research method widely used in health studies in recent years (Hsieh & Shannon, 2005) and is becoming popular in other fields as well.

To answer the aim of this chapter, the focus is on Australia as a case study. The case study research strategy was considered to be the most appropriate strategy to address the aim of the chapter and provide a local context for the findings. As Yin (1994: 1) states. "In general, case studies are the preferred strategy when 'how' or 'why' questions are being posed, when the investigator has little control over events, and when the focus is on a contemporary phenomenon within some real-life context". As this chapter is exploring issues of 'why' and 'how', a case study approach was deemed the most suitable (Yin, 1994).

Research method

This study adopted the desk method to collect the data and content analysis to analyse the data.

Data collection

Desk research is about tracking down useful existing published information. Desk research is standard in health science and market research but is now becoming common in many other fields as well (Punch, 2005). It is useful because sometimes it is difficult to collect data through interviews and surveys because of financial and time constraints. Desk research deals with secondary data or data which can be collected without the necessity of field work. Data for this study was gathered from government publications, government departments, the Australian Bureau of Statistics (ABS), online databases and the Internet. The focus of the data collection was construction industry activity and workforce profile, the impact of technology and material innovation on the workforce, and challenges for workers and stakeholders in addressing the incidence of the ageing workforce in the construction industry.

Data analysis

Both qualitative (literature review of recent publications, statistical data gathered from government publications and other occasional papers reports, academic papers, media, newspapers, magazines, and the Internet) and quantitative data (from the ABS and government reports) were analysed using qualitative content analysis. Publications by researchers and academics in this field were also analysed. The analysis was divided into three main themes, namely the profile of the workforce in the construction industry and its implications; challenges the construction industry is going to face due to the ageing workforce; and whether innovation in construction material and techniques will prevent a workforce shortage in the construction industry.

Results and discussions

This section reports on the findings from the desk research data collection and content analysis. The following three themes emerged from the content analysis: (i) a mis-match between the demand for and supply of workforce; (ii) other issues such as training, health risk, employer attitudes and risks associated with employing an aged workforce; and (iii) innovation of construction techniques and construction material. This section reports on the findings regarding these themes and finally presents a roadmap to address these issues.

Supply and demand

Typical supply and demand trends within the construction industry are cyclic in nature, consisting of short-term levels of construction performance and an increase in the application for construction services, resulting in a rise in apprentice intake (Cole, 1998). Toner (2003) found this to be partly influenced by the industry's high level of self-employment and temporary employment.

This section offers an overview of construction activity profile and the future demand for workforce, followed by a workforce profile and the supply situation in future.

Construction activity profile

Construction activity analysis is primarily based on the Australian Industry Group Construction Outlook Survey, which was conducted in March /April biannually in conjunction with the Australian Construction Association, the main industry body representing the nation's major construction contractors. The survey covered the responses of 100 companies employing approximately 60,000 people with a combined turnover of about $23 billion.

In 2014 the total construction activity was valued at $204.5 billion, in real volume terms. In 2014 engineering construction was the largest sector (56.1 per cent) of all construction activity, followed by residential building (26.6 per cent) and non-residential building (17.2 per cent) (Australian Industry Group, 2016). From 2012 to 2014, engineering construction declined by $16.7 billion as mining investment fell from its record high peaks and as the rate of mining-related heavy industrial, rail and port infrastructure construction dropped. Engineering construction's share of total work done rose from 47.2 per cent in 2010 to 61.6 per cent in 2012, before dipping to 56.1 per cent in 2014 with the passing of the peak in resources investment. The value of residential work rose by $6.5 billion through 2013 and 2014 in response to low interest rates, growing demand, and ongoing population growth. Non-residential building and commercial construction expanded by $1.4 billion over the same period, mainly reflecting large projects underway in Melbourne and Sydney. However, overall levels of non-residential building work done have remained relatively subdued in recent years, with heightened risk aversion and weak investor sentiment weighing heavily on the sector.

Owing to the construction industry's diverse range of products and services, and its traditional structure of sub-contracted and legally licensed trade specialisations, few individual businesses grow to a size that commands a substantial market share. As such, the industry is overwhelmingly comprised of small businesses with fewer than 20 employees (98.6 per cent of construction businesses). Indeed, 60 per cent of construction businesses are sole operators with no employees. The majority (82.2 per cent) of these small businesses operate in the trade services sector of the building industry that includes plumbers, electricians, plasterers and a myriad of other specialist building trades. Medium-sized businesses (employing between 20 and 200 employees) made up 1.3 per cent of the total number of businesses while medium to large businesses (employing 200 or more persons) accounted for just 0.1 per cent of the total (see Table 5.1). While the majority of companies participating in the industry are Australian-owned, the majority of tier one contractors are foreign-owned, if not foreign-controlled. Only 5.9 per cent of construction businesses generated more than $2 million of revenue in 2013–14, while the majority (60.8 per cent) generated revenue of less than $200,000.

The nation's leading non-residential companies are forecasting a modest recovery in 2017 (Australian Industry Group, 2016). This will also be supported by growth in the service and infrastructure sector as well as the construction of apartments. Non-resource infrastructure activity revenue derived from

Table 5.1 Distribution of businesses by employment size, June 2014

Employee range	Number of businesses	Share of businesses (%)
Non-employing	201,785	60.0
1–19	131,546	38.6
20–199	4,698	1.3
200+	197	0.1
Total	**338,226**	**100.0**

Source: Australian Industry Group (2015)

engineering construction projects is forecasted to pick up by 5.5 per cent in 2017 as new rail and road projects are undertaken. A more substantial lift will be in the telecommunication sector, also in line with the rising National Broadband Network (NBN)-related investment in 2017. In addition, the non-residential sector is expected to strengthen slowly over the forecast period. In 2015 it declined by 2.3 per cent, but in 2016 it picked up by 1 per cent, and it is projected that by 2017 it will grow by 4 per cent (AI Group and Australian Constructors Association, 2016). This growth will create a demand for construction workers across all sectors: residential, non-residential, service and infrastructure.

The total construction employment was expected to increase over the remainder of 2016 by 26 per cent in response to rising support work. Growth is expected across all categories of the construction sector. By June 2017, the total employment is projected to increase by 2.9 per cent. Growth will be mainly boosted by a 4.0 per cent increase in the number of on-site employees, with the figures of off-site and sub-contractors increasing by 2.6 per cent and 2.0 per cent respectively. There will be supply constraints because of the growth in workforce demand as major infrastructure works come on stream and shortages are expected to emerge again. A total of 44.4 per cent of the respondents reported either significant or moderate difficulty in restructuring skilled labour in the first six months of 2016.

In construction, there are many trades. The projected numbers of workers needed for various trades for 2019 are presented in Figure 5.1.

This section presents the growth of the residential and non-residential sectors which will have a high demand for construction workers. The next section looks into the workforce profile and analyses the demand for and supply of workforce

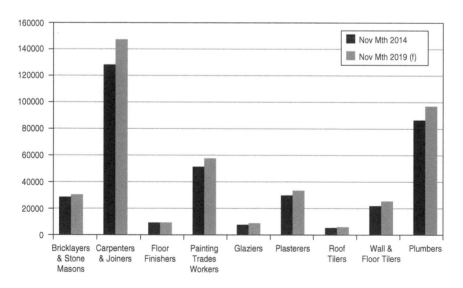

Figure 5.1 Construction trade projections

Source: https://sourceable.net/is-australia-heading-for-a-construction-trade-shortage/page3

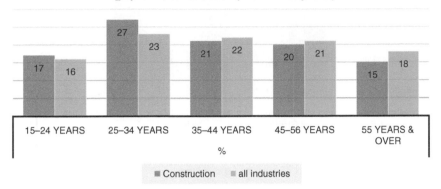

Age profile, construction (% of industry total)

Figure 5.2 Age profile
Source: Australian Government Department of Employment 2016, p14

because it is a well-established fact that the construction industry requires young workers owing to the physically demanding nature of the tasks involved.

Workforce profile (demographic pattern of labour force by age and trade projection)

The construction sector has a high percentage of workers aged 55 and above and is already facing serious skill and labour problems which are likely to increase in the coming years.

Figure 5.2 shows that the construction industry has a younger age group work-force compared to other industries. Around 35 per cent of the workforce is aged 45 years and over.

More than half of the workers in construction are technicians and trades workers. Consequently, the vocational education and training (VET) sector is a key supplier of skills. Around 52 per cent of the workers hold a certificate III or higher VET qualification and 54,000 young workers are employed as apprentices or trainees. Although the construction industry employs professionals such as civil engineers and architects, a mere 9 per cent of the total workers hold a bachelor's degree or higher qualification.

Over the six months leading to September 2016, the proportion of businesses experiencing difficulty in recruiting skilled labour was expected to remain at current levels (44.4 per cent having a significant or moderate challenge). However, difficulties for businesses in recruiting sub-contractors (50.0 per cent) were expected to become more widespread in the six months prior to September 2016.

Other issues

This section highlights three issues that impact the supply of workforce in construction industry, namely training, health and employer attitude to the aged workforce.

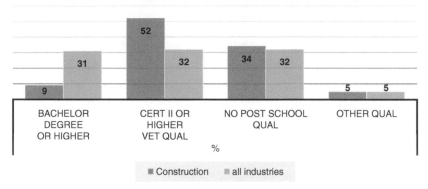

Educational attainment, Construction (% of industry total)

52

31   32   34   32

9   5   5

BACHELOR   CERT II OR   NO POST SCHOOL   OTHER QUAL
DEGREE   HIGHER   QUAL
OR HIGHER   VET QUAL

%

■ Construction   ■ all industries

Figure 5.3 Education profile
Source: Australian Government Department of Employment (2016: 14)

Education and training

There is the problem of low levels of qualification in the construction sector. Only 45 per cent of the workforce have formal qualifications and 29 per cent of employers provide structured training (Sundqvist et al., 2007). Watson (2012) contends that the Master Builders' Association has access to available training packages and they have found that the "…training system is struggling to keep up with the new demands of the building industry in response to changing building technologies and construction techniques, particularly in the highly competitive residential building sector" (Watson, 2012: 49).

Hampson and Brandon (2004) captured the views of construction employees in Australia on future innovations for the year 2020. They highlighted skills and education as high priorities and expressed concerns that skills are not being adequately developed to ensure the future of the construction industry. As Watson (2012) maintains, "…unless reforms are enacted to make training more attractive and accessible, the industry will face a growing shortage of skilled workers". The Housing Industry Association (HIA) highlights a persistent shortage of skilled trades workers in the residential construction industry (HIA, 2016). It has been found that for the apprenticeship system to continue, close attention must be paid to ensure that apprentices are offered a range of modern training practices (Toner, 2003). It is necessary for Australia to continue re-examining training and skill development packages to meet current industry demands.

As stated earlier, supply and demand are imbalanced. It was found that, in order to close the gap, there must be a greater emphasis on education and training.

Health risk

The ageing workforce in the construction industry has more complex health issues compared to workers in other industries because of the physically demanding nature of the occupation (Marchant, 2013). The main reason for the decline in health is biological decline. This decline may impact work because the workers rely on their physical capacity to carry out their jobs (Welch, 2009; Marchant, 2013). The common injuries or diseases affecting workers aged 55 and above include the loss of muscle mass and a subsequent decrease in strength; a decrease in bone density with an increase in age; musculoskeletal, back and neck complaints; knee problems; and a proneness to developing arthritis that will restrict their joint motions and functions (Hoonakker, 2010; Welch 2009). They are also prone to injury due to falling. As Shishlov et al. (2011) observed, the injury rate of construction workers aged 45 and over is twice as high as that for workers under 45. This poses a challenge for retaining the aged workforce in the construction industry.

Employers' attitude

Employer attitude is important for both retaining the aged workers and training new workers. In general, the majority of construction firms are small businesses with limited resources to support the training of workers and they do not like to contribute to support apprenticeships. Even though employers have positive attitudes towards the retention of the aged workforce, older workers often face discrimination (Schwatka et al., 2012).

Innovation of construction technology and material

It is useful to determine whether technological innovation could help reduce the need for labour in the construction industry. Advances in technology are closely related to the acceptance of alternative construction methods in the industry. New materials such as hebel wall panels and pre-fabricated/modular blocks are slowly becoming accepted and might reduce construction time on site, as well as cost and workforce requirements. Companies involved in alternative construction methods are utilising the existing trade workforce without the need for substantial additional training. Therefore, the impact on training a future workforce will be manageable (Construction Training Fund, 2015). The use of modular building material is currently at 2 per cent and expected to increase by 20 per cent over next 10 to 20 years (Construction Training Fund, 2015). The modular building material sector utilises subcontractors from the traditional industry sectors applying existing trade skills. Carpenters and wall and ceiling fixers are the most sought-after tradespeople and will continue to be in high demand even after new technologies and innovation in building materials come into play.

Construction work is physically intensive and this raises questions as to whether the older workforce is viewed as a burden or useful resource, and how the intake of young workers into the construction industry might be improved. Demographic changes to the population pointing to a decline in the number of people in younger age groups will lead to a number of challenges for the construction industry, particularly small firms. Such challenges include the health and fitness of their workers, an increase in the risk factor due to an ageing workforce, education and apprenticeship, material and technological innovation, and time flexibility. Producing more apprenticed workers in construction trades and enabling greater and more flexible involvement of older workers, as well as changes in material package weight and changes in policy may help overcome part of the problem. The next section presents the roadmap to address these issues.

Proposed roadmap

The results reflect the fact that the construction industry is a physically demanding industry and therefore more construction trade workers stop working at an earlier age than non-construction workers. Other countries (such as USA and Canada) are also being presented with the same trends of retirement within their construction industries (Welch, 2009). Usually, trade workers in this field start working from the ages of 17 to 19, so that by the age of 45 they have become physically worn out. Their health is adversely affected owing to the manual nature of the work itself. In addition, they become prone to back and neck problems, usually between the ages of 45 and 54. They are also likely to suffer from diseases such as arthritis. They may encounter joint movement issues or muscle and bone density problems. This phenomenon equates to a social capital loss within the construction workforce, and can also be directly linked to ageing. The hardships of construction work are cumulatively felt by both individuals and their teams, affecting overall productivity. The study also reflected that less up-skilling and training are taking place within the construction industry. Currently, there are insufficient numbers of people with the much-needed trades training. Linked to this, the CSIRO study (2016) predicted that by 2036, innovation in both technology and materials would reduce this industry's demand for physical work, but would increase the need for technology-linked assistance, which can be achieved through training workers in the optimal use of new technologies (Quezada, 2016).

Owing to these trends presented above, and the high levels of health risks that exist, there remains a need to retain workers aged 45 years and over. Reducing the physical demands within this field, regardless of age, is becoming essential. Many researchers have argued that this change will eventually be addressed by new technologies. To reiterate, the CSIRO study mentioned above and entitled 'What construction jobs will look like when robots can build things' (Quezada, 2016) has predicted that by 2036 innovation in technology and materials will reduce the physical nature of work in this field but increase the need for training in the use of new technologies. The government will also need to provide schemes to encourage people to easily obtain the minimum qualifications required, including safe workplace training, so that workers within this field are skilled (can read, write, and have much-needed business acumen). In turn, construction industry workers would have improved job opportunities, and their safety and longevity in the workforce might also be extended.

Emerging technologies offer new solutions for the construction industry. These technologies are improving processes while enabling automation. This perception can be justified after analysing digital collaboration platforms such as building information modelling (BIM), robot machine prototypes such as fast-brick robotics and the rapid progress of 3-D printing capabilities (Quezada, 2016; Erdogan et al., 2010). These innovations will eventually need more people skilled in the use of software programs and fewer people for labour-intensive jobs such as bricklaying or paving. Future technologies will elimi-nate dangerous and difficult tasks, particularly regarding the ageing workforce. However, this scenario will also increase the demand for training, and for more highly skilled labour. Governments will need to improve the training sector accordingly by improving existing incentives and financials associated with these training programmes or by working in partnership with industry. There is a need for governments to review apprenticeship training schemes to make training more rewarding and aligned to continued working opportuni-ties so that these pathways are appealing to young tradespeople (also known as *tradies*).

The construction industry needs both to invest in and produce alternative products (that limit hard labour), so that young tradies are not exposed to heavy tasks that will affect their health or wear them out at a young age. While keeping safe and healthy, the young trades-person or worker can be retained for a longer time. Such an intervention will, in turn, reduce many health problems such as back stress (in masons and other workers). This may merely entail adjusting work height or aligning mechanical support properly before undertaking works such as drilling (Welch, 2009).

Studies have also reflected that the pay negotiations in Australia between an employer and employee are aligned to both the national agreement and union. Work flexibility has been seen as a positive move in the residential sector because most of the work is being sub-contracted. It has also been noted that apprentices in Australia earn less than what they would earn from seeking social benefits, and therefore young unemployed persons do not find apprenticeships appealing. The government needs to set a more attractive salary for these apprenticeships, and this needs to be an amount that is much more than is offered by social security or the dole, in order to encourage young people to want to participate, or learn-by-doing in the construction industry. The government also needs to review the apprenticeship assistance accordingly, and to provide more money to the youth who undertake such difficult yet rewarding work. There should be more flexibil-ity for young apprentices to be able to negotiate with their employers in the commercial sector of construction. Currently, such negotiations are ruled by the unions, with much work progress and work culture being influenced by existing group dynamics. In addition, the government should review the minimum educa-tional requirements of Australian citizens in order to increase their employment opportunities: that is, the government should consider whether school-leavers have the basic levels of education required or whether additional training is required to be incorporated into the school-leavers' curriculum.

It has also been observed that employers from both the non-residential and service sectors prefer young workers rather than aged workers because of the levels of productivity that need to be achieved within set timeframes. On the other hand, in the residential sector, since work is sub-contracted, contractors have no hesitation in involving older or aged workers. The construction industry needs to support their workers with new technologies to reduce their physical efforts and minimise health and safety risks. This will, in turn, improve their work environment and ability to use new technologies. Employers require the participation of workers, trade unions and contractors to ensure the development of an effective solution that will improve overall workability.

Therefore new pathways for the construction industry that require workforce strategies and policies should be developed and supported by both the government and the construction industry in order to avoid an otherwise impending workforce shortage, attract young workers and retain the older or aged workforce. This could be enabled through:

- improved government policies;
- proper consideration and awareness of the health risk associated with the work;
- better technological innovations and material packaging;
- strategies to improve training associated with new technologies and safety; and
- the improvement of employer attitudes and support.

To achieve the above target for the construction workforce, there is a need for strong collaboration between the government, construction industries and the research and development sector with specific responsibilities for each sector (see Figure 5.4).

Further research will be required to recommend the modality of improvement for the above factors in order to address the shortage of workforce in construction industries by attracting young workers and retaining aged workers.

Summary

This chapter aims to identify issues of skills' shortage due to an ageing workforce. The chapter also provides a conceptual roadmap to address issues that will most likely be faced by the construction industry within the next twenty to thirty years. The chapter has clearly demonstrated that the industry will encounter a demand for trained and skilled workers in the future owing to many construction projects springing up in the various sectors. Such projects in Australia involve the service and infrastructure, non-residential and residential sectors. Many Australian states have an agenda to improve roads, infrastructure and services.

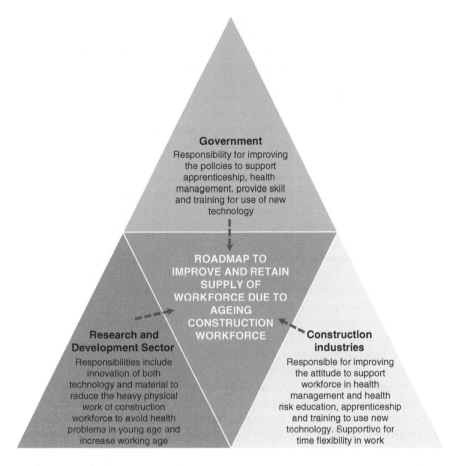

Figure 5.4 Proposed roadmap

And, in most of our states the residential market is still growing to meet expected future demands.

It can be observed that supply is not able to address the demand for labour in this sector because there is not enough labour available. There are two reason for this: firstly, the ageing workforce, and secondly, not many people are willing to work in the construction industry. The ageing workforce in the construction industry ranges from the ages of 45 and over, and this in itself is a very different demographic than that which is being encountered in other industries (since other industries have a much higher voluntary retirement age). It is clear that, owing to the intensively physical and demanding nature of the work, this industry is finding it difficult to retain its older workforce. Without automated facilities and modification of material packaging, an older and ageing workforce is going to find it increasingly difficult to remain in this field and undertake the challenging work required. The physically intensive work affects health detrimentally, with studies showing a decline from the ages of 45 and over.

To overcome these specific issues, there is a definite need for both government and industry to improve training and reduce health risks by automating the construction techniques, thereby ensuring flexibility in time, as well as to change employer attitudes. Also, there should be flexibility in negotiations between the employer and employee in the commercial sector of construction. Finally, government also needs to review the minimum level of education and training so that there will be a corresponding increase in employment opportunities.

The construction industry needs to find alternative products so that both young and middle-aged tradespeople are not exposed to lifting or unwrapping heavy materials or performing similar heavy tasks which cumulatively damage their health, and comprise their physicality. Maintaining the health and well-being of young workers should lead to their retention within the industry for a longer time. Therefore there is a need to increase research in the construction field.

The trend analysis shows that by 2036, the construction industry will have improved construction processes, enabling automation with the introduction of digital collaboration platforms, such as building information modelling (BIM) and robot machine prototypes such as the fast-brick robotics. These innovations will reduce the number of workers required, but workers will also need adequate training to use these new technologies.

To conclude, this chapter suggests that for the next 10 to 15 years there will be a shortage within the workforce because of the ageing population, but after that the gap between the demand and supply of labour may start diminishing. The gap will only vanish if workers are going to be sufficiently trained to use new technologies. Otherwise, the gap may remain due to the lack of adequate training (both of a technical nature and to deal with the cultural changes associated with utilising new technologies).

Finally, there is a need for both governments and the industry to work in partnership. A collaborative approach will ensure changes to: (i) reduce health risks; (ii) increase training and support; (iii) ensure a review of policies; and (iv) make construction work more appealing, safe and sustainable for younger workers.

References

ABS *see* Australian Bureau of Statistics.

AIGROUP & Australian Constructors Association (2016), Infrastructure to lift major project activity [Online]. Available at: http://cdn.aigroup.com.au/Economic_Indicators/Construction_Survey/2016/Construction_Outlook_June_2016_Final.pdf [Accessed 16 March 2017]

Alavinia, S. M., De Boer, A. G., Van Duivenbooden, J. C., Fringes-Dresen, M. H. & Burdorf, A. (2009), Determinants of work ability and its predictive value for disability, *Occupational Medicine*, Vol. 59, pp. 32–37.

Arber, S., Davidson, K. & Ginn, J. (2003), Changing approaches to gender and later life. In: S. Arber, K. Davidson & J. Ginn (eds), *Gender and ageing: Changing roles and relationships*, Open University Press/McGraw-Hill Educational, Maidenhead.

Armstrong-Stassen, M. & Templer, A. (2005), Adapting training for older employees: The Candian response to an aging workforce, *Journal of Management Development*, Vol. 24 (1), pp. 57–67.

Australian Bureau of Statistics (ABS) (2005), *Year Book of Australia* 1301.0, Australian Bureau of Statistics, Canberra.

Australian Bureau of Statistics (ABS) (2011), *Population benchmarks and labour force survey*, Cat 6291.0, Australian Bureau of Statistics, Canberra.

Australian Industry Group (2015), *Australia's construction industry: Profile and outlook*, July 2015, AiGroup Economic Research, Australia.

Australian Industry Group (2016), Construction outlook survey, *Construction Outlook*, June 2016, Australian Constructors Association, Australia.

Australian Industry Group Economic Research (2015), *Australia's construction industry: Profile and outlook,. July 2015*, AiGroup Economic Research, Australia.

Australian Physiotherapy Association (2016), *Ageing Australian workforce*, Australian Physiotherapy Association, Perth, Australia.

Balcombe, N.R. & Sinclair, A. (2001), Ageing: Definitions, mechanisms and the magnitude of the problem, *Best Practice Research Clinical Gastroenterology*, Vol. 15, pp. 835–849.

Becker, G. S. (1962), Investment in human capital: A theoretical analysis, *Journal of Political Economy*, Vol. 70(5), pp. 9–49.

Chartered Institute of Building (CIOB) (2015), *The impact of the ageing population on the construction industry*, The Chartered Institute of Building, Bracknell, UK.

Cole, L. J. R, (1998), *The Future of the Construction Industry 1998-2010*. JEASA Nominees Ltd. South Australia, Construction Industry Training Board, South Australia. 44.

Commonwealth Scientific and Industrial Research Organisation (CSIRO) (2015), *Are you ready for change? Farsight for construction. Exploratory scenarios for Queensland's construction industry to 2036*, CISRO, Canberra, Australia.

Construction Training Council (2014), *Construction Industry Workforce Development Plan 2014*, Construction Training Fund, Government of Western Australia.

Construction Training Fund (2015), The impact of new building techniques and technologies on the residential housing sector of the construction industry [Online]. Available at: https://bcitf.org/upload/documents/research_reports/WebversionImpa ctofNewtechniquesonResidentialSectorv20150110c.pdf [Accessed 17 March 2017].

Cook, S., Richardson, J., Gibb, A. G. F. & Bust, P. D. (2009), Raising awareness of the occupational health of older construction workers. *Paper presented at the CIB W099 International Conference, "Health and Well-being stream,"* Royal Melbourne Institute of Technology (RMIT), Melbourne Australia.

CSIRO *see* Commonwealth Scientific and Industrial Research Organisation

Department of Education Employment and Workplace Relations (2012), *Skills shortages by occupation and state or territory: Australian and New Zealand Standard Classification of Occupations (ANZSCO) 33 Construction Trades*, Labour Market Research and Analysis Branch, Department of Education Employment and Workplace Relations, Government of Australia, Canberra.

Department of Employment (2016), *Australian Jobs 2016*, Government of Australia, Canberra, Australia.

Eaves, S., Gyi, D. & Gibb, A. (2013), Building healthy construction workers by better workplace design: Understanding the context. In: S. D. Smith & D. D. Ahiaga-Dagbui (eds), *Proceedings of the 29th Annual Association of Researchers in Construction Management (ARCOM) Conference*, 2–4 September, Reading, UK.

Erdogan, B., Abbott, C. & Aouad, A. (2010), Construction in year 2030: Developing an information technology vision, *Philosophical Transactions of the Royal Society*, Vol. 368, pp. 3551–3565.

Frommert, D., Hofacker, D., Heien, T. & Andrea, H. J. (2009), Pension systems and the challange of population ageing: What does the public think? In: R. Edmondson & H. V. Kondratowitz (eds), *Valuing older people: A humanist approach to ageing*, pp. 139–158, The Policy Press, Bristol.

Garin, N., Olaya, B., Miret, M., Ayuso-Mateos, J. L., Power, M., Bucciarelli, P. & Haro, J. M. (2014), Built environment and elderly population health: A comprehensive literature review, *Clinical Practice and Epidemiology in Mental Health : CP & EMH*, Vol. 10, pp. 103–115.

Hampson, K. & Brandon, P. (2004), *Construction 2020: A vision for Australia's property and construction industry*, CRC Construction Innovation, Brisbane.

Hampson, K. & Brandon, P. (2006), *Construction 2020: A vision for Australian property and construction industry*, CRC Construction Innovation, Brisbane.

Heaton, A. (2015), Is Australia heading for a construction trade shortage? *Construction News*, 26 August [Online]. Available at: https://sourceable.net/is-australia-heading-for-a-construction-trade-shortage/ [Accessed 17 March 2017].

HIA *see* Housing Industry Association.

Hoonakker, P. & Van Duivenbooden, C. (2010), Monitoring working conditions and health of older workers in Dutch construction industry, *American Journal of Industrial Medicine*, Vol. 53(6), pp. 641–653.

Housing Industry Association (HIA). (2016), *HIA Trades Report*, HIA, Sydney.

Hsieh, H. -F. & Shannon, S. E. (2005), Three approaches to qualitative content analysis, *Qualitative Health Research*, Vol. 15(9), pp. 1277–1288.

Leaviss, J., Gibb, A. & Bust, P. (2009), *Understanding the older worker in construction*, Strategic Promotion of Ageing Research Capacity (SPARC), Reading, UK.

Lehtola, M. M., Van der Molen H. F., Lappalainen, J., Hoonakker, P. L. T., Hsiao, H., Haslam, R. A., Hale, A. R. & Verbeek, J. H. (2008), The effectiveness of interventions for preventing injuries in the construction industry, *American Journal of Preventive Medicine*, Vol. 35, pp. 77–85.

Manpower (2010), Strategic migration: A short-term solution to the skilled trades shortage, *World of Work Insight, Talent Shortage Survey*, Manpower, Sydney.

Marchant, T. (2013), Keep going: Career perspectives on ageing and masculinity of self-employed tradesmen in Australia, *Construction Management and Economics*, Vol. 31, pp. 1–16.

Mayring, P. (2000), Qualitative content analysis, *Forum: Qualitative Social Research*, Vol. 1(2), pp. 1–7.

McNair, S. & Flynn, M. (2006), *Managing an ageing workforce in construction: A report for employers*, Age Partnership Group, Department of Work and Pensions, London.

OECD *see* Organisation for Economic Co-operation and Development.

O'Hehir, J. (2014), *Age-friendly cities and communities: A literature review*, Centre for Work + Life, University of South Australia, Adelaide.

Organisation for Economic Co-operation and Development (OECD) (2015), *Ageing in cities*, OECD Publishing, Paris.

Plouffe, L. & Kalache, A. (2010), Towards global age-friendly cities: Determining urban features that promote active aging, *Journal of Urban Health*, Vol. 87(5), pp. 733–739.

Punch, K. (2005) *Introduction to social research: Qualitative and qualitative approaches*, Sage, London.

Quezada, G. (2016), What construction jobs will look like when robots can build things. *The Conversation*. [Online]. Available at: http://theconversation.com/what-construction-jobs-will-look-like-when-robots-can-build-things-63263.

Safe Work Australia (2015), *Construction industry profile*. [Online]. Available at: www.safeworkaustralia.gov.au/sites/swa/about/.../pages/construction-industry-profile

Schwatka, N. V., Butler, L. M. & Rosecrance, J. R. (2012), An aging workforce and injury in construction industry, *Epidemiologic Reviews*, Vol. 34(1), pp. 156–167.

Shishlov, K. S., Schoenfisch, A. L., Myers, D. J. & Lipscomb, H. J. (2011), Non-fatal construction industry fall-related injuries treated in US emergency departments, 1998–2005, *American Journal of Industrial Medicine*, Vol. 54(2), pp. 128–135.

Sivam, A. (2011), From NIMBY to WIMBY…Possibilities for housing options and new spatial arrangements of neighbourhoods to assist positive ageing in place. In: C. Whitzman (ed.), *Presented at the State of Australian Cities National Conference 29 November – 2 December*, University of Melbourne, Melbourne.

Sundqvist, J., Larsson, B. & Lindahl, G. (2007), Cooperation in the building sector between building material manufacturers and contractors to develop products. *The Australasian Journal of Construction Economics and Building*, Vol. 7(2), pp. 45–56.

Stemler, S. (2001), An overview of content analysis, *Practical Assessment, Research & Evaluation*, Vol. 7(17), pp. 1–6.

Toner, P. (2003), Supply-side and demand-side explanations of declining apprentice training rates: A critical overview, *The Journal of Industrial Relations*, Vol. 45(4), pp. 457–484.

Turnbull, J. A. B. (1990), *The elderly and the designed housing environment in Australia: Information, questions and guidance for planners, developers and architects*, Architectural Psychology Research Unit, Department of Architecture, University of Sydney.

United Nations. (2015), *World population prospects. The 2015 revision*, Working Paper No. ESA/P/WP.241, Department of Economic and Social Affairs, Population Division, United Nations, New York.

Van Dalen, H. P., Henkens, K. & Schippers, J. (2010), How do employers cope with an ageing workforce? Views from employers and employees, *Demographic Research*, Vol. 22, pp. 1015–1036.

Watson, M. (2012), Concerns for skills shortages in the 21st century: A review into the construction industry, Australia, *The Australian Journal of Construction Economics and Building*, Vol. 7(1), pp. 45–54.

Welch, L. S. (2009), Improving work ability in construction workers – let's get to work, *Scandinavian Journal of Work, Environment & Health*, Vol. 35(5), pp. 321–324.

WHO *see* World Health Organisation.

World Health Organisation (WHO). (2007), *Global age-friendly cities: A guide*, World Health Organisation, Geneva.

Yin, R. K. (1994), *Case study research: Design and methods*, Sage, London.

6 Indicative drivers of construction job insecurity in South Africa

Clinton Aigbavboa and Lerato Mathebula

Dare to know! Have the courage to use your own intelligence

(Immanuel Kant)

Background

In the past few decades, construction employees have increasingly been open/ exposed to the drivers of employment security (job security) as evident in many organisations (Borg and Elizur, 1992). These include mergers and acquisitions, closures, rapid restructurings, a general tendency to make organisations flatter, increasing use of non-fixed-term employment contracts, the advent of new technologies, and changes in the political landscape in several African countries (Låstad et al., 2015) which are often accompanied by large-scale layoffs. Also, the increasing turnaround of organisations and the requests for flexibility in work-life balance, along with the growing use of temporary employment, are interpreted to reflect a more insecure workforce (Låstad et al., 2015). Likewise, Schreurs et al. (2012) reportthat the recent global financial crunch aggravated the frequency and severity of changes in organisations to a large extent, thereby provoking increased feelings of job insecurity among employees.

The construction industry has been dealing with drastic changes over the past few years; changes that result in some negative consequences, one of these being job insecurity. These changes concern issues such as increased economic dependency among countries, rapidly changing consumer markets and escalated demands for flexibility within as well as between organisations (Sverke, Hellgren, and Naswall, 2006). However, construction organisations are responsible for the management of these changes to avoid issues such as job insecurity. Job insecurity in the construction industry can be defined as being uncertain about your employment status in a construction organisation owing to some reasons. The availability of jobs in the construction industry and related sectors has decreased above other industries, resulting in high unemployment and idle capacity.

In the seminal work of Greenhalgh and Rosenblatt (1984: 438), job insecurity (JI) is defined as the "perceived powerlessness to maintain desired continuity in a threatened job situation". Likewise, Hartley et al. (1991) suggested that the core of JI is the "concern about the future existence of the job". The general perception of JI is usually seen to include only discontinuation of employment or total job loss, but Hartley et al. (1991) and Borg and Elizur (1992) inform us that JI also includes a loss of popular job features which cannot easily be compensated for or prevented. Furthermore, according to Heaney et al. (1994) and Bernhard-Oettel et al. (2011), JI is an employee's overall concern about the continued existence of, or the threat to, his or her job. A general conclusion from the definitions above is that JI is based on perception and it is determined by individual differences across different organisational settings. Researchers have shown that the notion of JI is not necessarily linked with job instability or employment relationship instability (Ahn and Mira, 2002; Elman and O'Rand, 2002). Job insecurity is not about the fear or expectation of losing one's job per se. It is a personal conclusion regarding future potential job loss, and reduced confidence in future job stability (Heaney et al., 1994; Davy et al., 1997). Also, Bernardi et al. (2008) further extended this view by informing/indicating that JI is a subjective perception about employment conditions, specifically, about losing job stability/security and the continuity of a working relationship with an organisation. Hence, De Witte et al. (2012) stated that the perception of job insecurity is subjective, as the same objective situation can be interpreted differently by various employees. The personal nature of JI was also highlighted by Hellgren et al. (1999) when they distinguished between two different forms of job insecurity. These are quantitative job insecurity, which is a situation where employees worry about losing the job itself, and a qualitative job insecurity, which is the situation where employees worry about losing important features of the job, such as job stability, positive performance appraisals and promotions gains (Jacobson and Hartley, 1991; Ojedokun, 2008; Greenhalgh and Rosenblatt, 2010).

Job insecurity has been associated with adverse job-related outcomes. For instance, when faced with supposed JI, employees may report lower and reduced enthusiasm due to decreased job satisfaction and organisation commitment. Researchers reported that JI had been found to be a contributor that fosters a strong tendency or desire to resign from an organisation (Davy et al., 1997; Probst, 2000). In a meta-analysis, Sverke et al. (2002a, 2002b) found that job insecurity is negatively correlated with job satisfaction, trust, and job involvement, and positively correlated with employees' turnover intentions. Furthermore, the experience of JI is deeply affected by an employee's perception of the severity of an imminent adverse socioeconomic condition within the organisation to the extent that the inability to thwart/prevent the threat inflates the employee's sense of JI. In this context, scholars such as De Witte (1999) view JI as a major emotional stressor that reduces the well-being of the individual, increases physical strain and severely affects employee productivity (Cheng and Chan, 2008; Sverke et al., 2002b). Since the interpretation of JI can be influenced by sense making, this implies that JI has both individual and climate-level attributes. This means that JI can be conceptualised as individuals' perceptions of their situation, as well as their perceptions of the surrounding climate (i.e. employees sometimes worry about the future of their jobs at the workplace when the company appears not to be doing well). For instance, an employee may perceive the management's decision of organisational downsizing as threatening to his/her job continuation, whereas his/her colleague may regard the impact lightly. Likewise, Sverke et al. (2002b) inform us that the noteworthy outcomes of JI among employees are anxiety and stress. JI-induced stress leads to employees making decisions to leave/resign from their place of employment. This was further supported by the work of Alavi et al. (2013), who posited that JI, job dissatisfaction and the search for a new job are 'immediate consequences of employees' perceptions of a threat to their job continuation'.

Additionally, an insecure workforce can pose problems for construction companies as JI may cause employees to identify less with the company's corporate objectives and this may impact undesirably on motivation and readiness/ willingness to innovate and hence, on their individual productivity (Brockner et al., 1992; De Witte, 2005). Likewise, Ashford et al. (1989) inform us that the perception of JI has a direct relationship with employees' clear turnover intention, especially when many viable opportunities exist outside the firm. Also, Sutton (1983) and Greenhalgh and Rosenblatt (1984) posited that an increased intent by employees to leave a company as a result of JI leads to a situation where the most productive employees end up being the first to leave because of their high employability. This, in turn, results in reduced motivation and can subsequently jeopardise a company's productivity and competitiveness (Rosenblatt and Ruvio, 1996; O'Neill and Sevastos, 2013) in the long run.

Hence, the present study presents empirical findings of construction professionals' perceptions of the drivers of construction job insecurity in the South African construction industry with the aim of identifying the contributing factors and means to minimise the menace of job insecurity of construction workers. In the context of this study, construction workers' JI is defined as the perceived powerlessness to maintain desired continuity in a threatened job situation, as posited by Greenhalgh and Rosenblatt (1984). The conceptualisation stresses that JI is a perceptual phenomenon as informed by Sverke et al. (2002a). In other words, construction workers' JI is influenced by the subjective evaluation of employees.

Employment in the South African construction industry

South Africa's unemployment rate currently stands at 25 per cent (StatsSA, 2015), and one major goal that the country has set itself, according to the National Development Plan (The Presidency, 2012), is to drastically reduce the unemployment rate to 6 per cent by 2030. Studies have identified the construction industry as a major role player in job creation, not only in the construction sector but other areas of the economy as well. This is partly because the industry uses a broad range of inputs (such as construction materials) from other industries to produce its goods and services (UK Contractors Group, 2009). Thus, the industry contributes indirectly to the jobs that are created in these sectors. Examples of areas that benefit from construction output include manufacturing, mining, transportation, and real estate and business services. For instance, the cidb 2006 report estimated a direct job creation multiplier of around 4.2 jobs in the formal sector per R1 million invested ($142,045), and 2.3 employments in the informal sector per R1 million ($142,045) (CIDB South Africa, 2007). Additionally, it has been estimated that the construction materials manufacturing and materials distribution sector alone contributed to a direct job creation multiplier of around three (3) persons per R1 million ($142,045) in the country. The total direct employment multiplier in the construction and materials sector is estimated to be approximately nine (9) persons for every R1 million ($142,045) of investment (CIDB South Africa, 2015). Since 2008, construction has contributed about 9 per cent to total formal and informal employment in South Africa, while the contribution of construction to GDP has also been around 9 per cent (CIDB South Africa, 2015).

Furthermore, StatsSA (2014) revealed that the total construction works' spending in 2013 amounted to about R262 billion in 2010 Rands (or R310 billion in nominal Rands), and the industry currently employs approximately 820,000 people in the formal sector. The informal sector now accounts for a further 340,000 jobs (StatsSA, 2014). Overall, the construction industry is a significant employer of labour in South Africa, accounting for around 8 per cent of total formal employment and around 17 per cent of total informal employment (ProductivitySA, 2012).

Despite the laudable contribution of the South African construction industry to job creation, unemployment remains a serious concern in South Africa. Moreover, the construction sector has its particular difficulties: it utilises the fourth highest number of persons having no training, after agriculture, households, and mining. Considering the high rate of unemployment in the country, it can be expected that construction workers and professionals will be affected by job insecurity in their place of employment. However, Raiden and Dainty (2006) and Emuze et al. (2015) maintained that construction organisations show inadequate commitment to the development of people due to the belief that it is a costly function.

The distinctive features of the industry have implications for job security. For instance, when the overall construction climate is buoyant, few construction workers are unemployed, productivity increases and not many firms go bankrupt, as reported by Emuze et al. (2015). However, at low moments, Myers (2007) indicated that there is a reduction in construction activities, there are many unemployed workers as well as declines in production, and a significant number of firms are declared bankrupt. This clearly indicates that during times of economic crunch as the world has witnessed in the last decade, job security is also at its lowest owing to cutbacks in production. This will eventually lead to an increase in the number of layoffs, thus contributing to an increase in the unemployment rate of the country. The strong and growing focus on performance, emphasised/stringent deadlines for reaching targets as well as workplace accidents contribute to a climate of insecurity, if not to actual job losses and retrenchments in the construction industry. This aspect of the construction sector has fuelled a considerable amount of fear about job loss, especially amongst employees at lower levels. Hence, JI is a major concern in the construction sector.

Methodology

From the extant review of literature, 16 perceived causes of JI and 10 measures to minimise JI in the South African construction industry are defined in previous studies (see: Sutton, 1983; Greenhalgh and Rosenblatt, 1984; Davy et al., 1997; De Witte, 1999; Ahn and Mira, 2002; Elman and O'Rand, 2002; Låstad et al., 2015; Probst, 2000; De Witte et al., 2012). The identified variables were subsequently scored through the distributed questionnaire to the respondents. Having identified the contributing factors and ways to minimise JI through literature, an empirical survey was conducted among Gauteng Province construction professionals. Random samplings of the respective populations yielded a sample size of 50 small and medium-sized construction companies operating in the Gauteng Province of South Africa. From the identified firms, 132 construction professionals were surveyed using a mixed-mode quantitative survey. However, only 102 construction professionals completed the survey, equating to a 77 per cent response rate. This was considered satisfactory for the analysis based on the assertion by Moser and Kalton (1971) that the result of a survey could be seen as biased and of little value if the return rate was lower than 30 per cent to 40 per cent. Since the sample size for this study was relatively small, all groups of respondents were combined in the analysis to obtain significant results. Primarily, the principal objective of the survey, which forms part of the initial survey conducted as part of a larger Master's study, was to investigate the drivers of construction workers' job insecurity in the South African construction industry, as well as ways to minimise this.

The mixed-mode survey participants included architects, quantity surveyors, civil engineers and construction managers who are actively engaged in various construction projects undertakings in the Gauteng Province of South Africa. This yardstick was considered vital for the study to have a real understanding of the drivers of JI and ways to minimise it in the South African construction industry. The respondents were surveyed from July to August 2015. The questionnaire consisted mainly of closed-ended questions. The issues that were asked pertain to the extent to which individual factors contribute to construction workers' JI and to factors that could minimise construction workers JI. These closed-ended questions were based on a five-point Likert scale.

The data were analysed by calculating frequencies and the mean item score (MIS) of the rated factors. Although the empirical study is based on a relatively small sample of construction professionals, the findings provide an insight into the general perception of the drivers for JI and ways to minimise this menace in the South African construction industry. The calculation of the MIS is explained in the next section. The questionnaire was designed based on the information gathered during the literature review and does not form part of an existing survey instrument.

Mean Item Score (MIS)

A five-point Likert scale was used to determine the drivers of JI and likewise the factors that should be considered in minimising JI in the South African construction industry about the identified factors from the extant review of the literature. The adopted scale was as follows: (1) = strongly disagree; (2) = disagree; (3) = neutral; (4) = agree; and (5) = strongly agree. The five-point Likert scale scores were transformed to an MIS for each of the identified factors as scored by the respondents. The indices were then used to determine the rank of each item. These rankings made it possible to cross-compare the relative importance of the items as perceived by the respondents. The computation of the MIS was calculated from the total of all weighted responses and then relating it to the total responses on a particular aspect. This was based on the principle that respondents' scores on all selected criteria, considered together, are the empirically determined indices of relative importance. The index of MIS of a particular factor is the sum of the respondents' actual scores (on the five-point scale) given by all the respondents as a proportion of the sum of all maximum possible scores on the five -point scale that all the respondents could give to that criterion. Weightings were assigned to each response ranging from one to five for the responses of 'strongly disagree' to 'strongly agree'. This is expressed mathematically in Equation 1.0. The relative index for each item was calculated for each item as follows (Lim and Alum, 1995):

$$\text{MIS} = \frac{1n1 + 2n2 + 3n3 + 4n4 + 5n5}{\Sigma N} \tag{1}$$

Where: n1 = Number of respondents for strongly disagree; n2 = Number of respondents for disagree; n3 = Number of respondents for neutral; n4 = Number of respondents for agree; n5 = Number of respondents for strongly agree; and N = Total number of respondents.

Following the mathematical computations, the criteria were then ranked in descending order of their relative importance index (from the highest to the lowest). The next section of the article presents the findings of the survey and some discussion.

Results and discussion

The survey questionnaire was designed by asking respondents to identify the drivers that contribute to JI in the South African construction industry. The instrument further requested information about the ways in which the JI menace could be minimised. The respondents were able to identify the drivers as well as the means to minimise JI using a five-point Likert-type scale as described above.

Drivers of job insecurity in the South Africa construction industry

Table 6.1 presents the ranking of the drivers of JI in the South African construction industry. Table 6.1 indicates the respondents' perceptions of the drivers by

Table 6.1 Perceived causes of job insecurity in the South African construction industry

Perceived causes of job insecurity	MIS	SD	RANK
Lack of skills, experience and education	4.34	1.02	1
Unpaid overtime	4.28	1.07	2
Downsizing	4.28	1.07	2
Discrimination	4.18	0.98	3
Unequal treatment	4.18	1.06	3
Lower wages	4.18	1.12	3
Gender inequality	4.08	1.07	4
Decreasing salary	4.02	1.10	5
Unfair dismissals	4.02	1.24	5
Physical intimidation	3.98	1.08	6
Layoffs	3.60	1.05	7
Technological changes	3.42	1.01	8
Privatisation	3.38	1.16	9
Company restructuring	3.30	1.04	10
Merging	3.22	0.89	11
Acquisition	3.20	0.93	12

identifying the major factors that contribute to JI in the South African construction on a scale of 1 (strongly disagreed) to 5 (strongly agreed), and a mean score ranging between 1.00 and 5.00. It is worth noting that the mean scores were above the midpoint score of 3.00, which indicates that in general the respondents can be deemed to perceive the listed factors as contributing significantly to the problem of JI in the South African construction industry.

Based on the respondents' responses and the ranking of the weighted average from the MIS of the listed results in Table 6.1, it was found that the five major drivers of job insecurity amongst the 15 listed factors were a lack of skills, experience and education with an MIS of 4.34, which was followed by the issues of unpaid overtime, downsizing, rising costs, discrimination, unequal treatment and lower wages, gender inequality, decreasing salaries and unfair dismissals. Further causes included dismissals (MIS = 3.98 Rank = 6), physical intimidation (MIS = 3.60 Rank = 7), layoff (MIS = 3.42 Rank = 8), technological changes (MIS = 3.38 Rank = 9), privatisation (MIS = 3.30 Rank = 10), restructuring (MIS = 3.22 Rank = 11) and merging (MIS = 3.20 Rank = 12).

These findings accentuate/emphasise documented results of research that affirm that the first five listed factors are the major causes of JI in the construction industry (Caplan et al., 1980; Jeon and Shapiro, 2006; Sverke et al., 2006). Within the South African construction industry, the prevalence of the transformation of working life commonly known as downsizing which includes layoffs, merging, and restructuring that brings about insecure working conditions is a major cause of JI. The need to address these factors is thus vital in the South African construction industry, to curb reportedly challenging construction productivity-related problems.

Table 6.2 Measures to minimise job insecurity

Measures of minimising job insecurity	MIS	SD	RANK
Developing new skills and knowledge	4.72	0.61	1
Training employees (courses)	4.68	0.62	2
Developing employees' professional skills	4.66	0.56	3
Developing employees' capacities	4.54	0.61	4
Communication honestly with employees	4.38	0.75	5
Offering permanent work contracts	4.36	0.96	6
Implementing new task practices	4.30	0.79	7
Providing incentives to employee	4.18	0.83	8
Increasing organisational justice	4.02	0.98	9
Participating in company decision-making	3.94	0.91	10

Measures to minimise job insecurity in the South African construction industry

Table 6.2 shows the recorded acuities that pertain to the question that requested the participants to indicate on a scale of 1 (strongly disagreed) to 5 (strongly agreed) the extent to which the given factors contribute towards the minimisation of JI in the South African construction industry. From the table, it is worth noting that regarding the mean, only one MS is less than 4.00, which indicates that the other listed factors as adjudged by the respondents are deemed significant to have a major impact in the minimisation of JI in the South African construction industry. This means that the factors developing new skills and knowledge, training employees, communicating honestly, providing permanent work and offering contracts to employees are critical avenues through which JI can be minimised.

The current research findings are resonant with those of other scholarly works where it was found that the development of new skills, knowledge and training courses, for instance, are the most effective ways of minimising job insecurity (Batool and Batool, 2012; Sageer et al., 2012). Likewise, the work of Bhandari and Heshmati (2006) and Kuroki (2012) also concur with the research findings which found that the most useful measure to minimise job insecurity is to offer permanent work contracts to employees. However, the expansion of temporary employment in the construction industry is due to the limited lifespan of construction contracts, which increases the perception of job insecurity amongst existing workers in the industry as workers are laid off as soon as projects have been completed.

Summary

The study examined the factors contributing to the menace of job insecurity of construction workers in the South African construction industry as well as the means to minimise these. The investigation was conducted among construction professionals who are involved in various construction projects. The study acknowledges the limitation of the sample size used in the survey; however, the findings emanating from the study are indicative of the current trends on the subject of JI in the construction industry. Regarding implications, the study data confirmed that job insecurity is a complicated phenomenon which occurs when the employee is overwhelmed by fears concerning job continuation. The experience of JI is deeply affected by the individual's perception of the severity of negative socioeconomic conditions within the organisation. Also, the powerlessness to counteract the threat inflates/increases the employee's sense of JI. In this context, it is not surprising that some scholars view JI as a major emotional stressor that reduces the well-being of the individual, increases physical strain and severely affects employee productivity. Further implications emanating from the primary data confirmed that the prevalence of the transformation of working life commonly known as downsizing that brings about insecure working conditions is one of the leading causes of JI.

Hence, the study contends that the need to address these factors is thus vital in the South African construction industry to curb reportedly challenging construction productivity-related problems. When viewing the empirical results of the study, a further implication can be assumed to be that issues associated with human resources, capacity building of the construction, and information and documentation amongst workers are paramount to minimise JI in the South African construction industry. This further implies that if construction companies focus on empowering employees through obtaining the right skills, knowledge, and experience, paying workers the wages or salaries they deserve as well as promoting an equality policy amongst the workers, these measures can significantly reduce the concept/prevalence of job insecurity in the South African construction industry. Therefore, future studies should be directed towards how companies' strategic policies and human resources management processes can minimise the menace of job insecurity among construction workers.

References

Ahn, N. & Mira, P. (2002). A note on the changing relationship between fertility and female employment rates in developed countries, *Journal of Population Economics* 15(4): 667–682.

Alavi, S. Z., Mojtahedzadeh, H., Amin, F., & Savoji, A. P. (2013). Relationship between emotional intelligence and organizational commitment in Iran's Ramin thermal power plant. *Procedia - Social and Behavioral Sciences*, 84(0): 815–819.

Ashford, S. J., Lee, C., & Bobko, P. (1989). Content, causes, and consequences of job insecurity: A theory-based measure and substantive test. *Academy of Management Journal*, 32: 803–829.

Batool, A. & Batool, B. (2012). Effects of employees training on the organizational competitive advantage: Empirical study of Private Sector of Islamabad, Pakistan. *Far East Journal of Psychology and Business*, 6(1): 57–72.

Bernardi, L., Klaerner, A., & von der Lippe, H. (2008). Job insecurity and the timing of parenthood: a comparison between Eastern and Western Germany. *European Journal of Population*, 24(3): 287–313.

Bernhard-Oettel, C., De Cuyper, N., Schreurs, B., & De Witte, H. (2011). Linking job insecurity to well-being and organizational attitudes in Belgian workers: The role of security expectations and fairness. *The International Journal of Human Resource Management*, 22(9): 1866–1886.

Bhandari, K.A. and Heshmati, A. (2006). Wage inequality and job insecurity among permanent and contract workers in India: evidence from organized manufacturing industries. *Institute for the Study of Labor*, Paper no. 2097.

Borg, I., & Elizur, D. (1992). Job insecurity: Correlates, moderators and measurement. *International Journal of Manpower*, 13: 13–26.

Brockner, J., Grovner, S., Reed, T. F., & DeWitt, R. L. (1992). Layoffs, job insecurity, and survivors' work effort: Evidence of an inverted-U relationship. *Academy of Management Journal*, 35(2): 413–425.

Caplan, R. D., Cobb, S., French, J. R. P., Van Harrison, R. & Pinneau, S. R. (1980). *Job demands and worker health*. Ann Arbor: University of Michigan, Institute for Social Research.

Cheng, G. H.-L., & Chan, D. K.-S. (2008). Who suffers more from job insecurity? A meta-analytic review. *Applied Psychology: An International Review*, 57: 272–303.

cidb South Africa (2007). *The Building and Construction Materials Sector, Challenges and Opportunities*. Construction Industry Development Board, 2007. www.cidb.org.za

cidb South Africa (2015). *cidb SME Business Condition Survey; Quarter 3*. Construction Industry Development Board, October 2015. www.cidb.org.za

Davy, J. A., Kinicki, A. J., & Scheck, C. L. (1997). A test of job security's direct and mediated effects on withdrawal cognitions. *Journal of Organizational Behavior*, 18: 323–349.

De Witte, H. (2005). Job insecurity: Review of the international literature on definitions, prevalence, antecedents and consequences. *South-African Journal of Industrial Psychology*, 31(4): 1–6.

De Witte, H., De Cuyper, N., Elst, T. V. & Van den Broeck, A. (2012). The mediating role of frustration of psychological needs in the relationship between job insecurity and work-related well-being. *Work & Stress: An International Journal of Work, Health & Organizations*, 26(3): 252–271.

Elman, C. & O'Rand, A. O. (2002). Perceived job insecurity and entry into work-related education and training among adult workers. *Social Science Research*, 31(1): 49–76.

Emuze, F. A., Mputa, S. & Botha, B. (2016). A phenomenological study of candidate professional development in South African construction. *Journal of Construction*, 8(3): 1–7.

Emuze, F. A., Mputa, S., and Botha, B. (2015). A phenomenological study of candidate professional development in South African construction. *Journal of Construction*, 8(3): 1–7.

Greenhalgh, L., & Rosenblatt, Z. (1984). Job insecurity: Toward conceptual clarity. *Academy of Management Review*, 9: 438–448.

Greenhalgh, L. & Rosenblatt, Z. (2010). Evolution of research on job insecurity. *International Studies of Management and Organization*, 40: 6–19.

Hartley, J., Jacobson, D., Klandermans, B., & van Vuuren, T. (1991). *Job insecurity: Coping with jobs at risk*. London: Sage.

Heaney, C. A., Israel, B. A. & House, J. S. (1994). Chronic job insecurity among automobile workers: Effects on job satisfaction and health. *Social Science & Medicine*, 38: 1431–37.

Hellgren, J., Sverke, M., & Isaksson, K. (1999). A two-dimensional approach to job insecurity: consequences for employee attitudes and well-being. *European Journal of Work and Organizational Psychology*, 8(2): 179–195.

Jacobson, D., & Hartley, J. (1991). Mapping the context. In J. Hartley, D. Jacobson, B. Klandermans & T. van Vuuren (eds), *Job insecurity: Coping with jobs at risk* (pp. 2–22). London, UK: Sage.

Jeon, Doh-Shin, and Shapiro, Joel. (2006). Downsizing and job insecurity, mimeo, Universitat Pompeu Fabra.

Kuroki, M. (2012). The deregulation of temporary employment and workers' perception of job insecurity in Japan. *ILR Review*, 65(3): 560–577.

Låstad, L., Berntson, E., Näswall, K., Lindfors, P., & Sverke, M. (2015). Measuring quantitative and qualitative aspects of the job insecurity climate. *Career Development International*, 20(3): 202–217.

Lim, E. C. & Alum, J. (1995). Construction productivity: issues encountered by contractors in Singapore. *International Journal of Project Management*, 13(1): 51–58.

Myers, M. (2007). *Managing human resource development: a strategic learning approach*, 4th edition. Durban: Lexus Nexus.

Moser, C. A. & Kalton, G. (1971). *Survey methods in social investigation*. Oxford, UK, Heinemann Educational.

Ojedokun, A. O. (2008). Perceived job insecurity, job satisfaction and intention to quit among employees of selected banks in Nigeria. *African Journal for the Psychological Study of Social Issues*, 11(1): 204–219.

O'Neill P. & Sevastos, P. (2013). The development and validation of a new multidimensional Job Insecurity Measure (JIM): An inductive methodology. *Journal of Occupational Health Psychology*, 18(3): 338-349.

Probst, T. M. (2000). Wedded to the job: Moderating effects of job involvement on the consequences of job insecurity. *Journal of Occupational Health Psychology*, 5: 63–73.

ProductivitySA (2012). *Productivity Statistics for South Africa for 2012*. ProductivitySA. ISBN 978-0-620-58652-8.

Raiden, A. B. and Dainty, A. R. J. (2006). Human resource development in construction organisations: an example of a "chaordic" learning organisation?, *The Learning Organisation*, 13(1): 63–79.

Rosenblatt Z. & Ruvio A. (1996). A test of multidimensional model of job insecurity: the case of Israeli teachers. *Journal of Organizational Behavior*, 17: 587–605.

Sageer, A., Rafat, S. & Agarwal, P. (2012). Identification of variables affecting employee satisfaction and their impact on the organisation. *A Journal of Business Management*, 5(1): 32–39.

Schreurs, B. J., Hetty van Emmerik, I., Günter, H., & Germeys, F. (2012). A weekly diary study on the buffering role of social support in the relationship between job insecurity and employee performance. *Human Resource Management*, 51(2): 259–279.

Stats SA (2015). *Quarterly Labour Force Survey; July 2015*, Publication P0211. Statistics South Africa, www.statssa.gov.za

Stats SA (2014). *Quarterly Labour Force Survey; P0211, 4th Quarter 2014*. Statistics South Africa. www.statssa.gov.za.

Sutton, R. I. (1983). Managing organizational death. *Human Resource Management*, 22: 391–412.

Sverke, M. & Hellgren, J. (2002a). The nature of job insecurity: understanding employment uncertainty on the brink of a new millennium. *Applied Psychology*, 51 (1): 23–42.

Sverke, M., Hellgren, J. & Näswall, K. (2002b). No security: a meta-analysis and review of job insecurity and its consequences. *Journal of Occupational Health Psychology*, 7(1): 242–264.

Sverke, M., Hellgren, J. & Näswall, K. (2006). Job insecurity: a literature review. *National Institute for Working Life*. Stockholm, Sweden: Stockholm University.

The Presidency (2012). *National Development Plan 2030: Our future – Make it Work*. National Planning Commission, South Africa. Available at: https://nationalplanning-commission.wordpress.com

UK Contractors Group (2009). *Construction in the UK Economy: The Benefits of Investment*. Available at: www.ukcg.org.uk

Part II

A dissection of workplace diversity

7 A new approach to studying gender in construction

Natalie Galea, Adam Rogan, Megan Blaxland, Abigail Powell, Louise Chappell, Andrew Dainty and Martin Loosemore

Two things awe me most, the starry sky above me and the moral law within me
(Immanuel Kant)

Background

In the last few years, ethnography has become an increasingly popular approach in construction management research. Ethnographic research offers a powerful approach to getting 'under the surface' of construction practices enabling them to be reframed in ways "which account for both specificities of the context to which they relate and the socialities and materialities and experiences through which they unfold" (Löwstedt 2015: 404). Building on examples of early ethnographic research (Groat and Wang 2002; Davey and London 2005), more recent research has explored the application of ethnography to issues such as class, safety, trust, project complexity, architectural practice, intercultural communication and change management (see for example, Pink et al. 2013; Shipton and Hughes 2013; Subbiah 2012; Tutt et al. 2013; Thiel 2012; Oswald et al. 2014; Raisbeck 2016; Cummins et al. 2016; Orstavik and Dainty 2016). While some are experimenting with auto-ethnography as a way of providing an insider's point of view of the construction industry (Grosse 2016), most construction ethnography studies follow the exemplary ethnographic process. Some of these studies raise gender issues. For example, work by Sykes (1969), Applebaum (1981), Mars (2005), Paap (2006) and Theil (2007) reveal something of the 'masculinity' within which construction practice seems so enmeshed. Unlike our study, their focus has not been explicitly on gender.

The increasing use of ethnography in construction management research has generated discussion concerning its value and significance, as well as the various challenges that arise from such an approach. For example, Sage (2012) and Löwstedt (2015) argue that construction researchers are seldom self-reflexive, leaving the position and identity of the author unknown and unexamined regarding how their status, background, and experience influence the research process. Furthermore, 'traditional' ethnographic approaches involving a researcher being immersed in the activities of a certain field for an extended period present significant challenges when applied to the construction industry (Pink et al. 2013). The specific qualities and characteristics of construction workplace environments regarding limited time, spatiality, and flux often result in a range of restrictions that make long ethnographic studies challenging (Marshall and Bresnen 2012). Thus, recent research has called for new innovative approaches to ethnography that are more accessible and better suited to the shifting and temporal landscapes found in construction (Pink et al. 2013; Löwstedt 2015).

Drawing on the experiences of recent case study research into the formal and informal gendered practices of two multinational construction firms in Australia (Galea et al. 2015), this chapter explores the practical challenges of classic ethnography in the construction sector and how alternative ethnographic practices might offer pragmatic ways of overcoming them. Specifically, it shows how 'rapid' ethnography was adopted as a viable alternative and how this might be suitable to other social research in the broader field of construction management. It contributes to ongoing debates around the role of ethnography in construction research and demonstrates how rapid ethnography can be used effectively to examine the informal rules, norms, and practices that undermine gender equality in construction.

Importantly, the rapid ethnographic approach employed in this research was developed in line with recent work by Pink and Morgan (2013) on the value of short-term ethnography. Rather than seeing this as a "quick and dirty" variant of traditional ethnography (Hughes et al. 1995, 61), Pink and Morgan (2013) argue that short-term ethnographic research, characterised by intense and theoretically-informed research interventions produces focused and valid ways of knowing. While some early examples of rapid ethnography attracted criticism for being "superficial" (Knoblauch 2005) and separated theory and method (Manderson and Aaby 1992), Pink and Morgan (2013) maintain that ethnographic research that evolves in consistent dialogue with theory can produce alternative and intense routes to knowledge. As such, the focus of our approach was one of analytic intensity designed to foster both speed and rigour – drawing on research methods and practices used in rapid ethnography but always in dialogue with our theoretical framework – feminist institutionalism.

Reconceptualising gender equality through feminist institutionalism

As evidenced in previous chapters in this volume, the construction industry is one of the most male-dominated in Australia, despite efforts to address female representation (EOWWA 2012). Of 19 industry sectors in Australia, the construction industry performs the poorest, with women comprising only 11.4% of the workforce (ABS 2015), down from 17% in 2006 (ABS 2006). Women are also leaving the construction professions almost 39% faster than their male colleagues (APESMA 2010). Early enthusiasm from women construction professionals about their careers decreases with increased exposure to the construction workplace (Sang and Powell 2012; Dainty et al. 2000). The Australian construction sector is not unique. The proportion of women working in the UK construction sector has remained relatively unchanged since the mid-1990s, at 2% of construction trade workers and approximately 10% of professionals (Dutton and Wooley, 2015). This contrasts with the increasing number of women across the UK labour market, where employment rates for women have increased steadily over the same period. To explain these trends, we have turned to scholarship on gender in organisations and institutions. According to gender theorist Raewyn Connell (2005), gender is embedded in all organisations. Organisations, including those in the construction industry, are not gender-neutral structures but institutionalise practices of femininity and masculinity into unequal gender hierarchies, gendered cultures, and gender-specific jobs. In the construction sector, institutionalised gender practices play out through gender biases and discrimination in both formal policies and through practices and norms that serve to uphold a masculine workplace culture. This culture emphasises and rewards normalised masculinised behaviour, including presenteeism, total availability and long work hours (Powell et al. 2009; Watts 2007; Sang et al. 2014).

The theoretical lens used to understand women's underrepresentation within construction remains conceptually narrow. Dainty et al. (2007) note that construction management research is disconnected from theoretical developments in social and behavioural sciences and Terjesen et al. (2009) have identified an urgent need for more scholarship in gender research. There is clearly need to advance the intellectual debate in this area and to this end, new institutionalist theories, especially feminist institutionalism (FI), offer a valuable new lens (Chappell 2006; Waylen 2009; Mackay et al. 2010; Krook and Mackay 2010). The basic premise of new institutionalism is that rules – both formal and informal – 'matter' (March and Olsen 1984); they structure social interaction and decision-making in organisations by defining acceptable behaviours and sanctions imposed if these behaviours are not complied with (Helmke and Levitsky 2004). Rules are also a product of human agency and are constructed through processes of negotiation, conflict and contestation between different organisational interest groups (DiMaggio and Powell 1991). Importantly, they also exist within wider organisational, cultural, social and political environments and tend to reflect the wider institutions in those environments (Powell 2007).

FI is a new institutionalist approach that examines how institutions operate within organisations to maintain gender power imbalances and dynamics and shape organisational outcomes (Krook and Mackay 2010; Lowndes 2010, 65; Chappell and Mackay 2017). The value of the FI approach, in particular, is in the context of examining gender in organisations. FI provides a framework to examine hidden informal organisational rules, norms and practices that operate in conjunction, in conflict or in place of formal rules aimed at improving gender equality. According to new-institutionalism, formal rules are embedded in policies, procedures, initiatives, contracts and operational guidelines and are enforced through channels widely accepted as official (Lowndes 2005). In contrast, informal rules, norms, and procedures are unwritten, socially shared rules which are created, communicated, and enforced outside of officially sanctioned channels (Helmke and Levitsky 2004). In other words, actors identify that enforcement mechanisms exist if they were to break a rule or if they were to adhere to the rule. Chappell and Waylen (2013) and Azri and Smith (2012) argue that research using new institutionalist theory has prioritised formal rules; paying limited attention to informal rules. In large part, this is because informal rules are often taken for granted and tend to "shy away from publicity" (Lauth 2000, 26). But as FI points out (Krook and Mackay 2010; Chappell and Mackay 2017), the interplay of formal and informal rules is critically important to explaining the failure of formal policies to shift the gender imbalance in the political realm. We argue here that this interplay is equally as relevant in the construction sector. Examples of informal rules and practices with gender effects in the construction industry that act to undermine formal gender diversity policies include embedded views about what it takes to be a legitimate construction professional (Styhre 2011a; 2011b), including long hours, total availability and presenteeism (Lingard and Francis 2004; Watts 2007). These accepted norms of behaviour, and many others, keep the industry's dominant culture intact (Dryburgh 1999) and as a consequence act as a barrier to women's recruitment, retention, and progression.

The use of new institutional theory to understand how gendered power dynamics operate and are enforced and maintained in construction requires new methodological thinking (Dainty et al. 2007). The enforcement processes of formal rules can be more readily studied using traditional research methods since they are written down and involve obvious actors such as managers, policy makers and committees. However, informal rules are much harder to research since they are rarely written down and are often enforced and enacted through 'subtle, hidden, and even illegal channels' (Helmke and Levitsky 2004; Chappell 2006). Furthermore, analysing informal rules around gender practices raises practical issues as respondents may not even perceive or recognise the existence of informal gender rules because their normalisation and taken-for-granted nature may render them invisible.

Increasingly in other fields, these challenges are being addressed through ethnographic research (Radnitz 2011; Chappell and Waylen 2013). In simple terms, ethnography is a set of methods that involves the researcher participating in the daily lives of people for an extended period, observing what happens, listening to what is said, asking questions and collecting any relevant data that can throw light on issues of interest (Hammersley and Atkinson 1995). In ethnography, the researcher sees respondents as meaning makers and uses meaning-orientated methods to act as a translator between the group or culture under study and the reader (Millen 2000). By becoming a member of a community and engaging in its practices, a researcher can genuinely understand and uncover practices which would remain indigestible to an outsider (Brown and Duguid, 2000). For this reason, ethnography is typically more intensive than other forms of social research and its outputs are more holistic, descriptive and reflective in nature (Ybema et al. 2010). It also tends to avoid causal relationships in favour of an iterative-inductive approach (O'Reilly 2005), typically on a limited number of case studies that are investigated intensively in both a highly person-alised and field-based context using qualitative methods. This enables the researcher to capture the social meanings and ordinary activities of people in their natural settings. As such, ethnography must be seen as a reflexive and subjective practice within which the researcher is expected to contribute or participate and to openly recognise and acknowledge what they bring to the research setting (Pink et al. 2013). In this sense, ethnographers produce accounts which are both 'emic' (from the perspective of the respondent) and 'etic' (from the perspective of the researcher).

The use of ethnography to reveal gendered dimensions of social life has a long history (Skeggs 2001; Visweswaran 1997), but as Chappell and Waylen (2013) note, ethnography has rarely been used in new institutionalist research in specifi-cally gendered ways, or as a gendered lens on informal institutions. Similarly, while ethnography's use in construction management is on the increase, its value in exploring gendered rules and practices in the sector has yet to be fully explored.

Australian construction industry case studies

Motivated by the need to address the narrow theoretical and methodological understandings of the intransigence of gender inequality in construction, we adopted an ethnographic approach to investigate the interplay of informal gendered rules and formal rules and policies in two multinational construction firms. Company A is a privately owned multinational contractor, which operates in the commercial, residential, engineering and infrastructure markets. Company B is a publicly listed multinational contractor which operates in the commercial, resi-dential, engineering and infrastructure markets. The structure of both companies is typical and representative of large construction companies around the world.

This research intended to use a traditional ethnographic approach, with researchers charting developments in the field over an extensive period. However, we encountered some problems before starting our ethnography which challenged our approach. One concern was the highly time pressured and resource intensive nature of construction, which meant that research partici-pants had little time to engage with researchers. It was apparent that traditional ethnography, which would involve spending extended periods of time on site, would be too great an imposition on our respondents and compromise the qual-ity of data.

Second, the project-based nature of construction activity, across multiple loca-tions and for finite periods of time, meant traditional ethnography was not appropriate to investigate how informal rules interact with formal rules across different site locations and teams. It also became evident that it would not be possible to spend sufficient periods of time at various site locations. Accounting for the multiple temporalities which characterise construction practice has been acknowledged as representing a significant challenge for classic ethnographic approaches (Pink et al. 2010).

Third, our access to companies was through HR departments based at head office, who often acted as gatekeepers in accessing potential respondents. Our gatekeepers, who were usually the champions of gender-related initiatives, are especially concerned that project site teams, who were responsible for the day-to-day implementation of policies and initiatives, would be reluctant for us to observe and shadow them. However, interviews with business leaders in an earlier phase of the research suggested that while this may be the case for some, others on the site supported our fieldwork.

A more general concern that emerged early in the research related specifically to our research focus on gender. It quickly became clear that in both companies, *gender* was often understood to mean 'women' (Galea et al. 2015). We wanted to ensure that we were able to observe 'everyday' interactions, processes and work-place practices, not just issues understood to explicitly relate to gender, such as the provision of 'family friendly' work practices or women's leadership training. This was compounded by management concerns that men in the companies may think we were trying to 'catch them out' or identify them as the main barrier to women's recruitment, retention, and progression. A further concern was that men might be disinclined to participate in research about gender because, in thinking it was primarily about women, they did not consider that it related to them. We also faced a challenge with our university ethics approval process, where our deliberate effort *not* to mention gender was rejected due to lack of transparency and concern about misleading participants as to the intent of the research.

These issues forced us to reflect on the practicality of doing ethnography in the context of our research. Our experience reinforces Pink's (2005) observation that obtaining a richness of knowledge within a limited time frame using ethnography is challenging for researchers. However, to uncover the informal rules at work in shaping gender policy in practice, we were mindful of the need to maintain an ethnographic methodology. Our solution to this dilemma was to explore a 'rapid' ethnographic approach, long used in other fields such as medicine where researchers have employed 'time-deepening strategies' into 'rapid assessment procedures' to address similar problems (Harris et al. 1997).

Rapid ethnography as a solution to studying gender in the Australian construction sector

In contrast to the time-intensive approach of traditional ethnography, rapid ethnographers work in teams to undertake short, intensive investigations using multiple and iterative methods (Millen 2000; Isaacs 2013). In rapid ethnography, open-ended interviews and explorative observations are replaced with condensed equivalents, more focused on issues of interest which are identified from existing theory and literature before the research begins (Baines and Cunningham 2013). Broad conversations and interactions with numerous random informants are replaced with targeted and deliberate interviews with sampled respondents at key intervals and moments. In contrast to traditional ethnography which is highly explorative, this requires careful up-front planning directed by research which has a strong theoretical focus and is informed by specific research questions.

While rapid ethnography has been criticised by some for being a 'quick and dirty' approach to ethnography (Hughes et al. 1995; Millen 2000; Isaacs 2013), it provides a practical solution to the challenges we encountered, which would have otherwise compromised the quality of our data. As such we engaged a rapid ethnographic approach and incorporated strategies to address some of the challenges. As Pink and Morgan (2013) note, it is imperative that rapid ethnographic approaches are based on consistent and rigorous dialogue between theory and method. As such, the focus of our rapid ethnographic approach was one of analytic intensity designed to produce alternative and intense routes of knowing.

To address some of the concerns of our partners, the ethnographic phase of the research was repositioned and ultimately depoliticised, by shifting the focus away from gender per se, towards a proxy: *career pathways*. While for ethical reasons noted above information provided to participants acknowledged our interest in gender, we foregrounded our focus on recruitment, progression, and retention. This addressed concerns that telling people we were researching gender would limit our exposure to a broad range of people within the companies, particularly men. Gender, therefore, became an *analytical* concept used in our analysis, rather than the defining concept for the participants who we engaged in the research.

Given the challenges described above, the participant observation was also designed as a staged process, beginning with less intrusion and demands on individuals, to allow the companies and participants to become familiar with the research and to see that it would not interfere with their work. This strategy was adopted to build trust between the researchers and gatekeepers. The first part of the observation focused on pairs of researchers observing a range of company events and activities (14 in total). These included diversity training, new employee inductions, graduate assessment centres, training specific to leadership and construction skills, mentoring initiatives, management 'road shows', and diversity-specific events including a gender diversity strategy planning day.

Observations at these events included event locations, room layout and seating arrangements, presenters, timing of events, who attended, who had 'voice', tone of conversation and participant involvement and engagement, practices (who does what) and group dynamics (how do people participate etc.) and narrative (what is the message being reinforced). An observation template, which incorporated all these observation headings, was used as a prompt for researchers making notes in the field. Each event was attended by an insider and an outsider (both from the research team) to ensure in-built reflexivity and to incorporate different researcher observations and perspectives. The insiders had extensive experience in the construction sector, while the outsider researchers were gender experts from sociology and political science. This combination allowed the researchers to overcome challenges associated with being both an outsider – potentially not understanding issues – and an insider – missing important messages because they are taken for granted (Baines and Cunningham 2013; Bjarnegard 2013).

Our research team comprised of both men and women and, while not always possible, we actively tried to maintain this mix during fieldwork. As Pink et al. (2013) observe, construction is a masculine space, and much of the ethnography work to date has been undertaken by male researchers. This 'twinning' approach in which research teams were made up of both insider/outsider and male/female researchers had the effect of allowing different researchers to access differing types of interactions and responses. For example, insider researchers were able to engage and communicate using language and terminologies commonplace within the construction sector, yet unknown to outsiders, leading to more in-depth discussion and greater rapport. Similarly, male researchers in this study were given increased access to homosocial interactions between men at company events and on site, and male participants were more likely to discuss issues relating to sexuality and relationships with men as opposed to women. Furthermore, female researchers were more likely to be provided with in-depth responses from men regarding emotion and mental health. An example of the kind of differences experienced in the field by insider/outsider and male/female researchers is demonstrated in this excerpt from a debriefing interview between researchers:

> Female researcher: And regarding reverse sexism?
> Male researcher: Yep. So nearly every bloke I spoke to gave that narrative about women in construction.
> Female researcher: I never get this narrative. That's probably because they find out I was a woman in construction, maybe.
> Male researcher: Or maybe just the fact that you're a woman. I don't think they would say that to a woman because there's too much political incorrectness inherent in what they're saying.
>
> (Debrief Interview, Event 5, Company B)

As in the above example, in addition to the researchers making fieldwork notes of their observations, research pairs verbally debriefed at the end of each event. These debriefs were recorded and transcribed and became part of our data for analysis. This allowed the researchers to compare what they saw and experienced, to share how well they saw the theory fit with the case (or not) and to reflect on the method, including the different treatment depending on sex, placement, and other factors.

Where appropriate the researchers participated in the events they were observing and engaged in conversation with attendees asking questions such as 'Is this event/activity typical?' 'Is it important to attend events like this?' The researchers also left business cards so that they could be contacted confidentially after the events by any respondents who did not feel comfortable engaging in a discussion at the time. However, this strategy had limited success.

An example of an event observation is provided below. In this excerpt from a debriefing interview between researchers, the researcher is describing an observation at a leadership training event with a particular focus on group dynamics, power of speech, and gender performance:

> Regarding voice, who had a voice, it was fascinating ...the majority of the time men had the voice regarding the participants, but the voice of the leadership training people; [they] were women's voices. Concerning who was heard, when they were having open discussions within the room that was distinctive: a few men who repeatedly contributed. They were quite hyper-masculine guys too, very straight-talking with no qualms about swearing in front of the group. When they went into group discussions, even though there was probably one woman on each table, they were pretty even I thought regarding politeness and how they listened to each other. Certain groups had people who guided the conversation and that was really noticeable in the group participation exercise where each group who presented this new idea got up and presented for about 20 minutes each, and ... in every single one of those presentations, a man led and often it was the same man who finished and recapped each presentation. The women were all third or fourth speaker, so further down the line, and didn't have much voice within those presentations.
>
> (Debrief Interview, Event 4, Company A)

The event observations were successful in building trust within the companies that our research would not be intrusive and that the insights gained would be useful for the companies. This also involved debriefing with our company gatekeepers after some events about the process of the observation and our initial findings.

After the event observations, we conducted fieldwork at six construction project sites, one regional project, one medium sized metropolitan project, and one large metropolitan project for each of the two companies. These involved two researchers spending 3 to 5 days (depending on the size of the project) on site, observing, shadowing and interviewing professional employees in a range of positions. As outlined in Table 7.1, we shadowed 44 participants (36 men and eight women) and interviewed 81 construction professionals (49 men and 32 women).

To address concerns around confidentiality, we emphasised that we were observing process not content. Observations on site focused: on work practices (what time people arrived and left the site); roles on site (who does what, whether roles are associated with particular work practices such as total availability or leadership); whether there is a demarcation between project site and site office; the composition of work practices during the day (formal and informal meetings and interactions); who had 'voice' within these meetings and the tone of engagement; group dynamics (how do people participate); and narrative (what messages are being reinforced). The shadowing of participants provided a way to gather rich and abundant observations of how gender and gendered performance plays out within construction environments. For example, in the following

Table 7.1 Ethnographic research conducted and gender ratio

Research	Company A	Company B	Male	Female	Total
Event observations	7	7	–	–	14
Employee interviews	31	30	37	24	61
Participant observation: regional site	6	4	8	2	10
Participant observation: metropolitan site (medium)	8	8	13	3	16
Participant observation: metropolitan site (large)	10	8	15	3	18
Participant observation: total	24	20	36	8	44

accounts we see clear examples of how expressions of femininity and masculinity are enacted and performed in construction:

> So, as [Name, Female] and I made our way to the site where we had a meeting with the sub-contractor at 9.00am, it was a cleaning sub-contractor, I observed [Name, Female] pull her hair out, her mid length blonde hair, and sort of ruffle it out of the ponytail, and then place her hard hat on her head as she approached the site. This was quite an interesting statement or practice because I'd always pulled my hair in a bun or pulled it back up when I wore a hard hat because of the filth and the dust that would collect in my hair. I wondered whether she was performing femininity and whether she did this on purpose, and if so, what was the desired response? As a woman myself, I made an effort, a concerted effort to stand back really, and I mean a good 20 to 25 metres from [Name, Female] as I walked through the site. That way I could observe the impact she had on the site… and there was a lot of reaction. There was head-turning. Men checked her out. There was noise and chatter as she approached an area. I even saw one guy stick his tongue out – like a Maori does when they do a Haka – as she walked by… she was like a willy-willy [whirlwind] moving through the site. Men congregated in the parts where she was, and there were subtle hints to come over to her.
>
> (Participant Observation, Female 'Insider' Researcher, Site 4, Company B)

> One of the things I saw in contrast on site to in the office was the lack of amped-up or – how I put it – try-hard masculinity that I saw from most of the guys who were university-educated site and project engineers. What I mean by 'amped-up' is the fact that the guys on-site who were doing the work didn't need to peacock around because they were physically labouring – there was nothing to prove there around their masculinity – whereas the ones in the office embodied that chest-puffing masculinity that I didn't see on-site. There was no need to puff your chest out if you've already got a shovel in your hand and you're dealing with shovelling slurry, or you're driving a backhoe, or whatever. You're doing something. Whereas [in the office],

it was just a completely different masculinity, it was a put-on show. It was a performance. It was performative.

(Participant Observation, Male 'Outsider' Researcher, Site 2, Company A)

Shadowing provided an excellent opportunity for informal conversations with participants and included questions such as 'Was that a typical site meeting?', 'Is it important to arrive on site at this time?', 'Who is looked up to on this site, Why?'. Participants were also invited to take part in an interview if the conversation became more personal. Interviews that formed part of the rapid ethnography were designed to complement the observations and explore how formal and informal rules interact, conflict or are reinforced concerning recruitment, progression, and retention. As such, interviews explored narratives around career history and recruitment, mentoring and networks, what kind of people do well in the company, what kept them in construction, promotion processes and work practices (such as work hours and work-life balance). An example of the type of responses given in these interviews is found in this account of a female construction professional highlighting the importance of forming strategic alliances with senior male management to aid career progression and promotion:

I suppose being a female as well in a male-dominated industry I feel like I have to fight a lot more for what I want. I have to be the aggressive one, and let the right people know, or determine who the right people are that need to know that.

(Employee Interview, Female, Site 5, Company B)

As in the event observations, where possible, research teams attending project sites involved a combination of men and women and an 'insider' and an 'outsider'.

During the observation, researchers made notes and collected artefacts (photographs of room layouts, corporate messaging posted on walls or issued to employees and examples of gendered space and work practices). For example, the photograph below of construction site graffiti clearly demonstrates the gendered nature of such spaces.

After each observation the researchers debriefed and reflected, recording their conversation, which was later transcribed and formed the first stage of data analysis. A final obligation in any ethnographic research is to reflexively consider how empirical insights are kept in dialogue with theory throughout the research process (Pink et al. 2013; Pink and Morgan 2013). To address this issue, researchers also kept notes on how they negotiated their positionality in the field and how this impacted their analysis of the research.

Reflections on rapid ethnography as practice

This research provides significant insight into the value of rapid ethnography as a viable alternative in the study of gender equality in the construction industry.

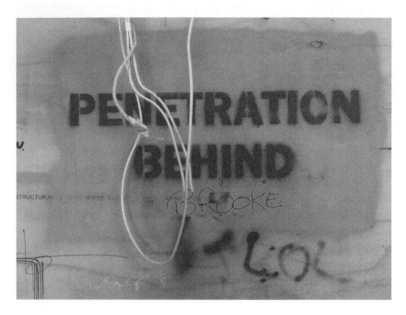

Figure 7.1 Gendered construction site graffiti

In adopting rapid ethnography to examine the interplay of informal gendered rules and formal rules and policies, this research revealed various advantages in such an approach. First, we were able to gather a large volume of high-quality, in-depth data that produced surprising and often deeply personal insights into the working lives of construction professionals. Second, this data was collected from multiple construction sites and event locations across a range of settings and contexts which enriched the value of our data. This allowed the research team to identify themes and issues that were shared across sites and locations which would not have been possible had we chosen a more traditional approach typically focused on one site. Third, our approach of using multi-person research teams, including both insider/outsider and male/female researchers, allowed us to gather different types of data and different perspectives. For example, insider researchers were able to establish rapport with participants by demonstrating knowledge of the industry and industry practices, whereas outsider researchers were able to direct questions at the taken-for-granted. Likewise, researchers found the responses given by participants often varied based on their gender. Fourth, our approach allowed for a consistent dialogue between theory and method as we moved through each stage of our research. This included consistent debriefing and reporting of findings to sharpen focus and refine research questions and practices throughout the research process.

While there were numerous advantages to conducting rapid ethnography, there were also some disadvantages. For example, there were significant logistical complexities in organising multiple site visits and event observations across the two companies. A traditional approach would typically involve a long-term engagement in a single site. The costs and time involved in collecting data across multiple locations at multiple times are somewhat higher than those found in traditional ethnography. There were some difficulties in building rapport and trust with participants 'rapidly' in the relatively short space of time spent on each site, though this was found to improve the longer researcher teams spent on the site. Having said this, researchers noted that the quality of the data from observations and interviews conducted with participants was much higher than from using interviews alone. We also found that the presence of the researcher had an impact on participants. Traditional ethnographic practice typically involves a single researcher spending a significantly longer period on the research site, which allows for developing trust and familiarity over time. In our research, multi-person teams conducted focused yet transient research over short periods of time, which ultimately made it harder for researchers to fade into the background.

Summary

As construction activity becomes ever more resource constrained and time pressured, the need for researchers to innovate in finding new ways to undertake social research which fits with the cultural and real-life constraints of their informants will no doubt grow. By addressing the practical difficulties of using ethnography in researching gender, we identified many benefits in adopting a more rapid approach to ethnography. These include less intrusiveness and avoiding the opportunity-cost of broad ranging unfocused investigations which would have resulted in the collection and analysis of a large amount of data which did not relate to the problem being addressed. However, to avoid criticisms of shallowness compared to traditional ethnography, rapid ethnography needs a strong theoretical context and systematic method to be thought out in advance. It also requires teamwork, close interaction with informants, focus, new technology such as digital media, multiple methods, up-front planning, multi-tasking, time and informant sampling and the identification of clear research questions in advance. Furthermore, it is imperative that such research maintains a comprehensive dialogue between empirical insights and theoretical framework throughout the research process.

Is rapid ethnography always appropriate? The answer is obviously no. But when faced with a range of practical constraints, rapid ethnography represents an interesting and potentially viable alternative to traditional ethnographic approaches, which fits with the cultural and practical constraints of the construction industry and which can provide meaningful empirical insights into the important and persistent problem of gender equality. It might also be of significant wider interest and value to many other aspiring social researchers in the field of construction management.

Acknowledgements

The authors would like to thank the participants and companies that took part in this research. This research was supported under the Australian Research Council's Linkage funding scheme (LP130100402).

References

ABS (2006), *Census of Population and Housing, Cat no 2068.0*. Canberra: Australian Bureau of Statistics.

ABS (2015), *Labour Force Australia, Detailed Quarterly, February 2015, Cat no 6291.0.55.003*. Canberra: Australian Bureau of Statistics.

APESMA (2010), *Women in the Professions: The State of Play 2009-10*. Melbourne: Association of Professional Engineers, Scientists and Managers.

Applebaum, H. (1981), *Royal Blue: The Culture of Construction Workers*. New York: Holt, Rinehart and Winston.

Azri, J. R. & Smith, J. K. (2012), 'Unwritten Rules: Informal Institutions in Established Democracies'. *Perspectives on Politics*, 10 (1): 37–55.

Baines, D. & Cunningham, I. (2013), 'Using Comparative Perspective Rapid Ethnography in International Case Studies: Strengths and Challenges'. *Qualitative Social Work*, 12 (1): 73–88.

Bjarnegard, E. (2013), *Gender, Informal Institutions and Political Recruitment: Explaining Male Dominance in Parliamentary Representation*. Basingstoke: Palgrave Macmillan.

Brown, J. S. & Duguid, P. (2000), *The Social Life of Information*. Harvard: Harvard Business School Press.

Chappell, L. (2006), 'Comparing Institutions: Revealing the 'Gendered Logic of Appropriateness'. *Politics and Gender*, 2 (2): 223–235.

Chappell, L. & Mackay, F. (2017) 'What's in a Name? Mapping the Terrain of Informal Institutions and Gender Politics'. In: G. Waylen (Ed.) *Gender and Informal Institutions*. New York: Rowman & Littlefield.

Chappell, L. & Waylen, G. (2013), 'Gender and the Hidden Life of Institutions'. *Public Administration*, 91 (3): 599–615.

Connell, R. (2005), 'Advancing Gender Reform in Large-Scale Organisations: A New Approach for Practitioners and Researchers'. *Policy and Society*, 24 (4): 5–24.

Cummins, A., Graham, B., Thomas, K. & Lucey, T. (2016), 'The Effectiveness of Managing Through Trust the Middle Management Layer of a Construction Company: Proposed Ethnographic Research'. In: P. W. Chan & C. J. Neilson (Eds.) *Proceedings of the 32nd Annual ARCOM Conference, 5–7 September 2016*, Manchester, UK, Association of Researchers in Construction Management, Vol 2: 991–1000.

Dainty, A., Bagilhole, B. & Neale, R. (2000), 'A Grounded Theory of Women's Career Under-Achievement in Large UK Construction Companies'. *Construction Management and Economics*, 18 (3): 239–250.

Dainty, A., Green, S. and Bagilhole, B. (2007), *People and Culture in Construction: A Reader*. London: Taylor and Francis.

Davey, L. M. J. & London, K. (2005), 'An Ethnographic Study of Knowledge Management Practices within an Australian Market Leader: An Organizational Culture Perspective'. In: F. Khosrowshahi (Ed.) *Proceedings 21st Annual ARCOM Conference, 7–9 September 2005*, London, UK, Association of Researchers in Construction Management, Vol. 2: 1237–46.

DiMaggio, P. & Powell, W. (1991), 'Introduction'. In: P. DiMaggio & W. Powell (Eds.) *The New Institutionalism in Organizational Analysis*. Chicago: University of Chicago Press, pp. 1–40.

Dryburgh, H. (1999), 'Work Hard, Play Hard: Women and Professionalization in Engineering – Adapting to the Culture'. *Gender and Society*, 13 (5): 664–682.

Dutton, M. & Wooley, G. (2015) *Diversity in Construction*. Edinburgh, UK: Edinburgh Napier University.

EOWWA (2012) *Women in the Workforce by Industry*. Canberra: Equal Opportunity for Women in the Workplace Agency, Australian Government.

Galea, N., Powell, A., Loosemore, M. and Chappell, L. (2015), 'Designing Robust and Revisable Policies for Gender Equality: Lessons from the Australian Construction Industry. *Construction Management and Economics*, 33 (5–6): 375–389.

Groat, A. & Wang, D. (2002), *Architectural Research Methods*. New York: John Wiley & Sons, Inc.

Grosse, H. (2016), 'An Insider's Point of View – Auto-Ethnography in the Construction Industry'. In: P. W. Chan & C. J. Neilson (Eds.) *Proceedings of the 32nd Annual ARCOM Conference, 5–7 September 2016*, Manchester, UK, Association of Researchers in Construction Management, Vol 2: 1001–1010.

Hammersley, M. & Atkinson, P. (1995), *Ethnography*. London: Routledge.

Harris, K., Norge J., & Stephen, F. (1997), 'Rapid Assessment Procedures: A Review and Critique'. *Human Organization*: Fall 1997, Vol. 56 (3): 375–378.

Helmke, G. & Levitsky, S. (2004), 'Informal Institutions and Comparative Politics: A Research Agenda'. *Perspectives on Politics*, 2 (4): 725–740.

Hughes, J., King, V., Rodden, T. & Andersen, H. (1995), 'The Role of Ethnography in Interactive Systems Design'. *Interactions*, 2 (2): 56–65.

Isaacs, E. (2013), 'The Value of Rapid Ethnography', In: B. Jordan (Ed.) *Advancing Ethnography in Corporate Environments: Challenges and Emerging Opportunities*. Walnut Creek, CA: Left Coast Press, pp. 92–107.

Knoblauch, H. (2005), 'Focused Ethnography'. *Forum Qualitative Sozialforschung/Forum: Qualitative Social Research*, North America, 6 Sep. 2005. Accessed January 24, 2013 (www.qualitative-research.net/index.php/fqs/article/view/20).

Krook, M. L. & Mackay, F. (2010), *Gender, Politics, and Institutions: Towards a Feminist Institutionalism*. Basingstoke: Palgrave Macmillan.

Lauth, H. J. (2000), 'Informal Institutions and Democracy'. *Democratization*, 7: 21–50.

Lingard, H. & Francis, V. (2004) 'The Work-Life Experiences of Office and Site-Based Employees in the Australian Construction Industry'. *Construction Management and Economics*, 22: 991–1002.

Lowndes, V. (2005), 'Something Old, Something New, Something Borrowed ... How Institutions Change (and Stay the Same) in Local Governance'. *Policy Studies*, 26 (3): 291–309.

Lowndes, V. (2010), 'The Institutional Approach'. In: D. Marsh & G. Stoker (Eds.) *Theories and Methods in Political Science*, Basingstoke: Palgrave.

Löwstedt, M. (2015), 'Taking Off My Glasses in Order to See: Exploring Practice on a Building Site Using Self-Reflexive Ethnography'. *Construction Management and Economics*, 33 (5–6): 404–414.

Mackay, F., Kenny, M. & Chappell, L. (2010), 'New Institutionalism through a Gender Lens: Towards a Feminist Institutionalism?' *International Political Science Review*, 31 (5), 578–588.

Manderson, L. & Aaby, P. (1992), 'Can Rapid Anthropological Procedures be Applied to Tropical Diseases?'. *Health Policy and Planning*, 7 (1): 46–55.

March, J. G. & Olsen, J. P. (1984), 'The New Institutionalism; Organizational Factors in Political Life'. *The American Political Science Review*, pp. 734–749.

Mars, G. (2005), 'Locating Causes of Accidents in the Social Organisation of Building Workers'. *International Journal of Nuclear Knowledge Management*, I (3): 255–269.

Marshall, N. & Bresnen, M. (2012), 'Where is the Action? Challenges of Ethnographic Research in Construction'. In: S. Pink, D. Tutt, & A. Dainty (Eds.) *Ethnographic Research in the Construction Industry*, Abingdon: Routledge, pp. 108–24.

Millen, D. R. (2000), *Rapid Ethnography: Time Deepening Strategies for HCI Field Research*. New York: AT&T Labs- Research.

O'Reilly, K. (2005), *Ethnographic Methods*. Abingdon: Routledge.

Orstavik, F. & Dainty, A. (2016), 'On the Doing of Building Work: 'Ways of Knowing' As Modes of Coping with Complexity'. In: P. W. Chan & C. J. Neilson (Eds.) *Proceedings of the 32nd Annual ARCOM Conference*, 5–7 September 2016, Manchester, UK, Association of Researchers in Construction Management, Vol 2: 707–716.

Oswald, D., Smith, S. & Sherratt, F. (2014), 'A Spanish Subcontractor in a UK Culture'. In: A. Raiden & E. Aboagye-Nimo (Eds.) *Proceedings of the ARCOM 30th Annual Conference*, 1–3 Sept 2014, Portsmouth, UK, Association of Researchers in Construction Management, pp. 259–268.

Paap, K. (2006), *Working Construction: Why White Working-Class Men Put Themselves and the Labour Movement in Harm's Way*. New York: Cornell University Press.

Pink, S. (2005), 'Dirty Laundry: Everyday Practice, Sensory Engagement and the Constitution of Identity'. *Social Anthropology*, 13 (3): 275–290.

Pink, S. & Morgan, J. (2013), 'Short-Term Ethnography: Intense Routes to Knowing'. *Symbolic Interaction*, 36 (3): 351–361.

Pink, S., Tutt, D. & Dainty, A. (2013), *Ethnographic Research in the Construction Industry*. Abingdon: Routledge.

Pink, S., Tutt, D.E., Dainty, A. R. J. & Gibb, A.G. (2010), 'Ethnographic Methodologies for Construction Research: Knowing, Practice and Interventions'. *Building Research and Information*, 38 (6): 647–659.

Powell, A., Bagilhole, B. & Dainty, A. (2009), 'How Women Engineers Do and Undo Gender: Consequences for Gender Equality'. *Gender, Work and Organization*, 16 (4): 411–428.

Powell, W.W. (2007), 'The New Institutionalism'. In: S. R. Clegg & J. R. Bailey (Eds.) *The International Encyclopaedia of Organization Studies*. Thousand Oaks, CA: Sage.

Radnitz, S. (2011), 'Informal Politics and the State'. *Comparative Politics*, 43 (3): 351–371.

Raisbeck, P. (2016), 'The Architect as Gleaner: Design Practice as Performance in the Architectural Office'. In: P. W. Chan & C. J. Neilson (Eds.) *Proceedings of the 32nd Annual ARCOM Conference*, 5–7 September 2016, Manchester, UK, Association of Researchers in Construction Management, Vol 2, pp. 809–818.

Sage, D. (2012), 'The Trials, Tribulations and Translations of an Ethnographic Researcher in Construction'. In: S. Pink, D. Tutt, & A. Dainty (Eds.) *Ethnographic Research in the Construction Industry*. Abingdon: Routledge.

Sang, K. J. C., Dainty, A. R. J. & Ison, S. G. (2014), 'Gender in the UK Architectural Profession: (Re)Producing and Challenging Hegemonic Masculinity'. *Work, Employment and Society*, Vol. 28 (2): 247–264.

Sang, K. J. C. & Powell, A. (2012), 'Equality, Diversity, Inclusion and Work-Life Balance in Construction. In: A. Dainty & M. Loosemore (Eds.) *Human Resource Management in Construction: Critical Perspectives*. Abingdon: Routledge, pp. 163–196.

Shipton, C. & Hughes, W. P. (2013), 'Making Changes in Practice: An Ethnographic Study of a Hospital Project'. In: S. D. Smith & D. D. Ahiaga-Dagbui (Eds.) *Proceedings of the ARCOM 29th Annual Conference*, 2–4 Sept 2013, Reading, UK, Association of Researchers in Construction Management, pp. 1113–1123.

Skeggs, B. (2001), 'Feminist Ethnography'. In: P. Atkinson, A. Coffey, S. Delamont, J. Lofland & L. Lofland (Eds.) *Handbook of Ethnography*, Thousand Oaks, CA: Sage, pp. 426-442.

Styhre, A. (2011a) 'In the Circuit of Credibility: Construction Workers and the Norms of "a Good Job"'. *Construction Management and Economics*, 29: 199–209.

Styhre, A. (2011b) 'The Overworked Site Manager: Gendered Ideologies in the Construction Industry. *Construction Management and Economics*, 29: 943–955.

Subbiah, A. (2012) 'An Investigation of the Factors Influencing the Success of Construction Planning for the 2012 Olympic Stadium: An Ethnographic Study'. In: S. D. Smith (Ed.) *Proceedings of the ARCOM 28th Annual Conference*, 3–5 Sept 2012, Edinburgh, UK, Association of Researchers in Construction Management, pp. 1035–1045.

Sykes, A. J. M. (1969), 'Navvies: Their Social Relations', *Sociology*, 3: 157–72.

Terjesen, S., Sealy, R. & Singh, V. (2009), 'Women Directors on Corporate Boards: A Review and Research Agenda'. *Corporate Governance: An International Review*, 17 (3): 320–337.

Thiel, D. (2007), 'Class in Construction: London Building Workers, Dirty Work and Physical Cultures'. *British Journal of Sociology*, 58 (2): 227–251.

Thiel, D. (2012), *Builders: Class, Gender and Ethnicity in the Construction Industry*. Abingdon: Routledge.

Tutt, D. E., Pink, S., Dainty, A. R. J. & Gibb, A. G. F. (2013), 'Building Networks to Work: An Ethnographic Study of Informal Routes into the UK Construction Industry and Pathways for Migrant Up-Skilling'. *Construction Management and Economics*, 31 (10): 1025–1037.

Visweswaran, K. (1997), 'Histories of Feminist Ethnography'. *Annual Review of Anthropology*, 26: 591–621.

Watts, J. (2007), Porn, Pride and Pessimism: Experiences of Women Working in Professional Construction Roles. *Work, Employment and Society*, 21 (2): 219–316.

Waylen, G. (2009), 'What can Historical Institutionalism Offer Feminist Institutionalists?'. *Politics and Gender*, 5 (2): 245–253.

Ybema, S., Yanow, D., Wels, H. & Kamsteeg, F. (2010), 'Ethnography'. In: A. Mills, G. Durepos & E. Wiebe (Eds) *Encyclopaedia of Case Study Research*. Thousand Oaks, CA: Sage, pp. 348–352.

8 Misplaced gender diversity policies and practices in the British construction industry

Developing an inclusive and transforming strategy

Linda Clarke, Elisabeth Michielsens, and Sylvia Snijders

Out of the crooked timber of humanity no straight thing was ever made
(Immanuel Kant)

Background

For hundreds of years the proportion of women in construction occupations has been so low that – except in times of acute manpower shortages – parity with men seems to remain elusive. Paradoxically, the statistical record reveals continuous presence of women in construction, which calls for constant examination of the structure of the industry, to understand and identify possible mechanisms of exclusion, and to seek ways to correct the imbalance. The obstacles to integration of women have been shown to include inappropriate and poor working and employment conditions, particularly long working hours, male-dominated training courses, discriminatory recruitment practices, based on word of mouth rather than qualifications, the persistence of a macho culture, lack of work–life balance possibilities, and short-term concerns with output (Fielden et al. 2000; Clarke et al. 2004; Clarke and Gribling 2008; Sang and Powell 2012). The lack of state regulation of employer responsibility, as well as the very nature of the labour market and training available, has also played an important role (Clarke and Wall 2014). Many of the same barriers, such as concerning recruitment and retention, are shared at both operative and professional levels, although there are some differences. As female participation at operative level is particularly low, this is the main focus here.

Addressing deep-rooted structural issues has represented a significant challenge for the industry. Policies and practices – largely founded on the rationale of the 'business case' – have been formulated, but little progress has been made over the last 30 years. Why is this so? This chapter seeks to answer this question, examining the policies and practices put forward to combat low gender participation, their focus, the case on which these are built, and the degree to which the structure of the industry in Britain is conducive to their implementation. Drawing on existing literature, extensive research of our own, and analyses of census, industry and WERS (Workplace Employment Relations Study) statistics, the relative irrelevance of a 'business case' for greater gender participation in construction is shown, given that this is primarily focused on persuading employers to take responsibility for changing the situation.

Allegiance to the 'business case' is particularly misplaced given the structure of the construction industry. This is particularly the case for the larger construction organisations, as they are becoming increasingly finance-driven and largely project-focused, reliant on extensive subcontracting chains, networks, agency labour, and the self-employed, directly employing only professional and technical staff, and withdrawing from membership of employers' associations. At the same time, the industry is dominated by small or very small firms, often acting as subcontractors, where they are rarely in a position to train, target recruitment, and adopt diversity policies and practices. It is as a result more and more difficult to identify who 'the employer' actually is, and, consequently, who the 'employee' is, as many are self-employed, and many more are working through agencies or consultancies.

The chapter argues that another strategy is required for the sector to become more inclusive, one not focused solely on a 'business case' and corporate social responsibility, but on involving policy-makers and employees in transforming a male-dominated industry into a gender-neutral industry. The chapter begins with an overview of female participation in construction in the United Kingdom (UK), and the nature of diversity management (DM) in the sector. Findings from three case study projects are then presented, followed by a discussion of the limitations of the industry in promoting gender inclusivity, and considerations for the future.

A changing situation for women in construction?

Of the 2,225,000 working in the UK construction industry, only 12% are women (ONS 2014), a proportion which has remained relatively stable over a 25-year period since 1990 (Briscoe 2005; The Construction Index 2012). Considering the different construction-related occupations, as shown in Table 8.1, representing the results of the 1991 and 2011 censuses for construction-related managers and professionals, the low representation of women is much more pronounced in the so-called 'manual' occupations, and in on-site roles (Lingard and Francis 2004). Women's presence at operative level was below 3% in 1991, and there have been hardly any positive changes in the ensuing two decades, and even some negative trends (see rows 18, 19 and 21); the increase in female participation in construction is almost solely at professional level (see rows 1 to 17) (Clarke and Wall 2014). Even at professional level, female integration is poor: women in 2011 still represented only 22% of architects, 10% of quantity surveyors, and 12% of managers and professionals, despite a 100% increase overall since 1991. The technical

Table 8.1 Changes in female participation in selected construction occupations 1991 to 2011

Row number	Occupation SOC90 1991[1]	% women 1991	Occupation SOC2011[2]	% women 2011	% change 1991-2011
	TOTAL all occupations	**44.04**	**TOTAL all occupations**	**47.09**	**3.06**
1	111 Managers in building & contracting[3]	5.80	1122 Production managers & directors in construction	11.60	5.80
2	210 Civil, structural, municipal, mining & quarrying engineers	2.12	2121 Civil engineers	8.37	6.25
3	211 Mechanical engineers	2.70	2122 Mechanical engineers	5.08	2.38
4	212 Electrical engineers	1.59	2123 Electrical engineers	3.20	1.62
5	213 Electronic engineers	2.66	2124 Electronics engineers	6.09	3.43
6	216 Design & development engineers	2.33	2126 Design & development engineers	7.70	5.36
7	217 Process & production engineers	3.38	2127 Production & process engineers	12.37	8.99
8	219 Other engineers & technologists nec*	7.35	2129 Engineering professionals nec*	13.56	6.21
9	260 Architects	10.77	2431 Architects	21.53	10.76
10	261 Town planners	21.83	2432 Town planning officers	39.59	17.76
11	312 Quantity surveyors	4.57	2433 Quantity surveyors	10.13	5.56
12	2180 Planning & quality control engineers	8.16	2461 Quality control & planning engineers	23.12	14.96
13	301 Engineering technicians	4.25	3113 Engineering technicians	8.80	4.54
14	304 Building & civil engineering technicians	11.20	3114 Building & civil engineering technicians	19.70	8.49
15	309 Other scientific technicians nec*	20.86	3119 Science, engineering & production technicians nec*	26.15	5.30
16	303 Architectural & town planning technicians	14.95	3121 Architectural & town planning technicians	27.99	13.04
17	310 Draughtspersons	9.86	3122 Draughtspersons	17.35	7.48
18	52 Electrical/Electronic Trades	2.66	52 Skilled metal, electrical & electronic trades	2.62	-0.04

	SOC90 1991		SOC2011		
19	537 Welding trades	5.74	5215 Welding trades	2.29	-3.45
20	5210 Electricians, electrical maintenance fitters	1.24	5241 Electricians & electrical fitters	1.72	0.47
21	5290 Other electrical/electronic trades nec*	3.29	5249 Electrical & electronic trades nec*	2.92	-0.37
22	50 Construction Trades	1.35	53 Skilled construction & building trades	2.28	0.93
23	535 Steel erectors	0.80	5311 Steel erectors	1.29	0.49
24	500 Bricklayers, masons	0.58	5312 Bricklayers & masons	1.41	0.83
25	501 Roofers, slaters, tilers, sheeters, cladders	0.81	5313 Roofers, roof tilers & slaters	1.62	0.82
26	532 Plumbers, heating & ventilating engineers & related trades	1.24	5314 Plumbers & heating & ventilating engineers	1.73	0.49
27	570 Carpenters & joiners	0.83	5315 Carpenters & joiners	1.21	0.38
28	503 Glaziers	2.15	5316 Glaziers, window fabricators & fitters	2.74	0.59
29	509 Other construction trades nec*	1.34	5319 Construction & building trades nec*	2.67	1.34
30	502 Plasterers	0.43	5321 Plasterers	1.18	0.75
31	506 Floorers, floor coverers, carpet fitters & planners, floor & wall tile fitters	0.86	5322 Floorers & wall tilers	2.03	1.17
32	507 Painters & decorators	2.61	5323 Painters & decorators	4.62	2.01
33	505 Scaffolders, stagers, steeplejacks, riggers	0.52	8141 Scaffolders, stagers & riggers	1.15	0.63
34	9230 Road construction & maintenance workers	1.04	8142 Road construction operatives	2.11	1.08
35	9290 Other building & civil engineering labourers nec*	0.93	8149 Construction operatives nec*	2.80	1.87

Source: own calculations based on ONS data for 2011 census data, and OPCSC (1994) for 1991 census data

[1] Occupation SOC90 1991 census 10% sample: All usual residents aged 16 or over in employment the week before the Census GB
[2] Occupation SOC2011 census 2011: All usual residents aged 16 or over in employment the week before the Census England/Wales
[3] Each category gives occupation name and number as listed in SOC90 or SOC2011
* nec = not elsewhere classified

occupations, where the proportion of women was 26% in 2011, and female participation had increased by 5.3% since 1991, fare much better, which is significant, given the growing importance of technical competencies, particularly to low-energy construction and to building information modelling (BIM).

While participation of women in the building trades is low, they have always maintained a presence, originally entering through apprenticeships; the earliest records confirming this date to the sixteenth, seventeenth and eighteenth centuries (Clarke 2007; Clarke and Wall 2014). Their numbers dwindled to less than 0.3% of the total skilled construction labour force by the end of the nineteenth century, as the trade unions and the engineering professions adhered to a policy of deliberately excluding women from entering either an apprenticeship or scientific education (Clarke and Wall 2009; Drake 1984). In the twentieth century, in particular during the two world wars, women were encouraged to take up jobs that became available as a result of conscription of men, resulting in over 25,000 women working in the construction industry in 1943, with a participation rate of 3.8% (Clarke and Wall 2011). This disproves the longstanding argument against women entering manual occupations based on their physical abilities; in times of labour shortage, when there are no other available sources of labour, women have been actively recruited to fill vacant positions, and they have performed the work successfully.

It was not until the 1970s and 1980s, after the Sex Discrimination Act (1975) became law, that a combination of political change and grassroots campaigning led to considerable numbers of women training and working in construction (Wall and Clarke 1996; Michielsens et al. 1997; WAMT 2001; Payne 1991). Local authorities, committed to changing their male-dominated construction workforce, created a framework of support for women in the trades, by providing designated women's officers, by holding regular meetings, by placing more than one woman on any site, by introducing flexible working hours, and through a clear and transparent set of equal-opportunities guidelines, backed up by internal procedures to address grievances. The success of these measures is evident in the presence of 266 women in manual construction occupations in just seven Inner London local authority building departments – known as direct labour organisations (DLOs) – in 1989. Leicester DLO continues this legacy, employing 123 women in 2012 as part of its 431-strong workforce, with 18 of the 75 craft apprenticeships held by women (Clarke et al. 2006; Craig and Oates 2014).

Today, although DM is established as good-practice policy in most sectors in the UK, this has not led to a significant improvement in female participation in construction, particularly at operative level. The suitability of DM measures to instil changes is anyway questionable, given that women may prefer to build reputation through hard work and 'fitting in', rather than relying on the support initiatives of those human resource (HR) departments that do exist in construction (Sang and Powell 2012; Francis 2015). Indeed, as is evident from Table 8.2, the management questionnaire (MQ) and the survey of employees questionnaire (SEQ) of the Workplace Employment Relations Study (WERS) show the

Table 8.2 Percentage of firms in construction and other industry sectors (SIC 2007)[1] adopting equal-opportunity policies and practices

Firms by size	Construction sector					All SIC 2007 industry sectors	
	S^2	S–M	M–L	L	All	% of firms from all sectors	% of firms from the best-performing sector
Formal equal-opportunity policies					86	90	100
Recruitment and selection practices							
Special procedures to encourage women to apply	4	8	19	18	11	12	21[3]
Monitor by gender	8	24	62	64	33	42	80[3]
Review for indirect discrimination by gender	8	24	37	46	27	35	68[3]
Promotion practices							
Monitor by gender	4	11	30	18	15	21	60[3]
Review for indirect discrimination by gender	8	8	30	36	17	29	54[3]
Working time arrangements							
Working flexitime (where an employee has no set start/finish time)	11	20	19	27	19	16	33[3]
Job-sharing schemes	0	0	4	9	2	7	21[4]
Reducing work hours (e.g. by switching from full- to part-time)	4	18	8	9	11	16	31[5]
None of the above	46	6	12	0	21	10	0[6]

Source: Own calculations, based on WERS MQ and SEQ (2011)

[1] Each category gives the industry sector as listed in the SIC 2007
[2] Each category of small (S), small to medium (S–M), medium to large (M–L) and large (L) was identified using the ONS Construction Statistics Annual Table
[3] The best-performing sector here is the public administration and defence sector
[4] The best-performing sector here is education
[5] The best-performing sector here is wholesale and retail
[6] The best-performing sector here is financial and insurance activities

superficiality of diversity measures in construction, as well as the continuing presence of what Hoque and Noon (2004) have termed 'empty shell' policies.

While the proportion of construction firms with formal DM policies found in the WERS survey of 2011, namely 86%, compares well with the all-industry average, namely 90%, there are sectors, such as public administration and defence, where 100% of firms' report having formal DM policies in place. These policies, too, do not necessarily translate into diversity practices in construction firms. Indeed, the construction sector does not always compare positively with the all-industry average; other sectors perform markedly better about recruitment, selection and promotion procedures. In public administration and defence, for instance, 80% of firms monitored reported having recruitment and selection procedures for gender in place, compared with only 8% of small firms in construction, and 33% of all construction firms (see Table 8.2). Furthermore, in the education sector, 21% of firms reported providing job-sharing schemes as working time arrangements, compared with none of the small and small-to-medium construction firms, and only 9% of the large ones (see Table 8.2). The majority of the 798 (766 males and 32 females) employees covered in the WERS survey as working in the construction sector also reported that flexible working time arrangements were not available to them, and, while more than half of these female employees reported that flexitime and a reduction in working hours were available, these working time arrangements were predominantly used in large construction firms. Size of firm, including self-employment, therefore has a significant influence on implementation of diversity practices.

Employer-led diversity policies and practices in construction – an 'empty shell'

Improving gender diversity through organisational policies and positive action in construction is clearly severely limited, because, as the data show, these policies and practices do not automatically lead to monitoring, inclusive practices, and increased gender participation. The appropriateness of the employer-led 'business case' approach, on which DM is based, is therefore open to question. The 'business case' rationale of DM relies on the business benefits that this case provides. It focuses on employer action, as first promoted in the early 1990s (Ross and Schneider 1992; Kandola and Fullerton 1998). Cox and Blake's (1991) original DM framework categorises benefits as external benefits and internal benefits, where external benefits relate to the organisational context, such as recruitment of 'best talent', compliance with legislation, and employer branding. In contrast, internal business benefits result from improved operations within the firm. The 'business case' argument for a better gender balance in the construction industry, however, is mainly linked to the external commercial benefits of employing women, which include the following: tackling industry skill shortages; increased profitability, inward investment, effectiveness and customer satisfaction; reduced likelihood of litigation, high staff turnover, high recruitment/training costs, high rate of absenteeism and loss of corporate knowledge/intellectual capital; and a more motivated, committed, and productive workforce (UKRC 2005; Bagilhole 1997; Barnard et al. 2010; Dainty et al. 2000). Compliance with legislation, in particular, is seen as an important driver of gender diversity considerations, given the potentially negative impact of bad publicity from a discrimination case (CIPD 2006, 2007; English and Le Jeune 2012). Nevertheless, the measure of whether these aspects have improved the internal workings of the firm remains an unknown.

Even though evidence of the positive impact of DM on performance, and indeed on equality and diversity, remains weak and mixed, and an ethical argument (linked to inclusion and justice) might be more apt, there is a widespread belief in the benefits of a diverse workforce for business (Dickens 1999; Kochan et al. 2003; Özbilgin and Tatli 2011; Wright et al. 2014). However, DM itself fails to challenge structural inequalities, because diversity is treated as a corporate benefit rather than a social right, which leads only to partial and voluntary management implementation. This reinforces an individual approach, which hinders collective participation of employees in the identification, assessment and adaption of diversity measures to specific contexts. In segregated contexts, such as construction, structural barriers to equality and diversity are particularly entrenched, and therefore in need of more concerted and appropriate actions than the 'business case' offers. Indeed, an employer-focused DM approach is particularly inappropriate and ineffective in UK construction, due to its particular structural characteristics, regarding firm size, self-employment, employee participation, employment and working conditions, and education and training.

Approach and methodology

To dissect the effectiveness of the 'business case', publicly available secondary data has been analysed, including from the 1991 and 2011 censuses and the 2011 Workplace Employment Relations Study (WERS) management questionnaire (MQ) and survey of employees questionnaire (SEQ). The WERS data were analysed about the implementation of diversity policies and practices, to ascertain in what kind of firm these policies and practices have relevance, and to what extent they are just an 'empty shell' (Hoque and Noon 2004). The WERS 2011 MQ data are representative of 2,680 workplaces with five or more employees in the UK, and the SEQ data are representative of 21,981 employees. In total, 103 construction firms were included in the MQ data, and 798 construction employees in the SEQ data, 4% of whom were women. Construction firms and employees were identified using the Standard Industrial Classification (SIC) 2007 major groups and the Standard Occupational Classification (SOC) 2010 major groups.

Based on this data, an assessment was made, drawing initially on the annual Construction Statistics of the Office for National Statistics (ONS 2016a), concerning to what extent the structure of the UK construction industry is compatible with diversity policies and practices devised in support of the 'business case'. This was further supported by findings from in-depth qualitative research previously carried out by the authors in relation to mega projects, including Heathrow Terminal 5 and the Olympic Park, and on more recent work, namely the Thames Tideway Tunnel (Clarke and Gribling 2008; GLA 2007; Clarke and Holborough 2011; Clarke et al. 2015). All of these projects involved in-depth interviews with key stakeholders, including employers, trade union representatives, HR personnel, equal-opportunities officers, and training providers.

Changing gender participation on mega projects

In the last decade, concerted efforts to include women in construction have focused on mega projects, such as Heathrow Terminal 5 (known as T5), the Olympic Park, Crossrail, and, more recently, the Thames Tideway Tunnel. The size, complexity and nature of major projects mean that they are often highly regulated and subject to scrutiny, opening up the possibility for a more inclusive employment policy, as well as a change in the composition of teams. Such projects have the advantage of highlighting practical as well as policy steps that can be taken to increase gender inclusivity, including those that may not necessarily have been very successful.

Heathrow Terminal 5 (T5)

T5 was one of Europe's largest construction projects of the time, costing £4.3 billion over five years, from 2003 to 2008, requiring about 8,000 workers at peak, and comprising 16 major projects and over 147 sub-projects (Clarke and Gribling 2008). The client, the British Airports Authority (BAA), took a proactive position on labour management issues, including consideration of diversity and equality concerns (BAA 2004). Trade unions played an important monitoring role, ensuring employee participation and maintaining good industrial relations. While the policy of direct employment was extremely successful in showing that there is an alternative to casual employment in construction, the diversity achievements on the T5 site were disappointing: only a few female electricians and one or two women in other occupations were employed in the operative workforce. Further, of the 150 employed over the course of three years, as a result of the local labour scheme, only 3% were women (Experian 2006). The local labour force became a dwindling minority, and the majority consisted of 'travelling' men, those classified as such in the collective agreement as eligible for an allowance for travel away from home. Many suppliers consequently came to rely heavily on an itinerant workforce, both from outside London and – increasingly – from different nationalities and geographic regions outside the UK. Few if any of the many hundred construction trainees in surrounding colleges, who included many women, found work on the site (Baker Mallett 2008).

The obstacles identified by Clarke and Gribling (2008) to obtaining a more inclusive and local labour force on T5 were the following: the lack of work experience and placements available allowing those from local colleges to gain practical skills in construction; the training on offer, which was largely confined to traditional trades and geared to domestic construction work; the means of recruitment, including reliance on agencies, which tended to target a traditional white male and migrant workforce; and the site working hours and shift patterns, which, coupled with the long trip to work, made the working day almost impossible for any but 'travellers' and migrants prepared to sign the working time directive and work intensively. This meant the site was almost structured to suit an itinerant rather than a local workforce, an aspect supported through the incentives given to 'travellers' in the pay structure.

The Olympic Park

The Olympic Park, which employed over 6,000 at any one time, and which modelled itself on T5, provides another example of the importance of securing an overriding agreement with all stakeholders, including contractors, subcontractors, trade unions, clients, and local authorities on working conditions, direct employment, and a preference for local residents (ODA 2011; Druker and White 2013). The responsibilities of the Olympic Delivery Authority (ODA) for securing employment and training and boosting skill levels locally were based on explicit targets for women, ethnic minority groups, disabled people, and local residents, including the requirement to place at least 2,250 people into trainee programmes, apprenticeships, and work placements. This meant monitoring and recording those employed – a difficult task given that over 800 firms were contracted to do the work (Foster 2010).

Despite all the efforts, according to London 2012 (2011), only 3% of operatives and 5% overall of those who worked on the Olympic Park were women, although the original target was 11% (ODA 2010). Employment of these women was largely attributable to the Women into Construction (WiC) project and positive action initiatives of the ODA, including the following: taster days for local women; supporting prospective employees with child and health care; targeting local colleges for female prospective recruits; on-the-job-learning for women on-site; and the Chicks with Bricks programme (EHRC 2011). The London Olympics was unprecedented in the targets for apprenticeships, underrepresented groups, and the establishment of the WiC, an organisation that is continuing today (Wright 2013, 2014a). Despite the disappointing results for women, the project provides a good example of the value of securing an overriding agreement with all stakeholders, of contract compliance, of setting equality targets, and, above all, of guaranteeing direct employment, which confirms the findings of the earlier study by the Greater London Authority on diversity in construction (GLA 2007).

Crossrail

A more recent example, focused particularly on improving professional gender participation, comes from the Crossrail project. This £14.8bn 26-mile (42-km) scheme has largely focused on employing women engineers, who currently constitute 10.7% of the membership of the Institution of Civil Engineers (ICE) and between 6% and 13% of engineers in the construction industry – the lowest figure in Europe (Kitching 2014). Driven partly by 'business case' concerns about a rapidly increasing number of engineering vacancies, and by assumptions that the more gender-balanced a team is, the better it performs (Wright et al. 2014; Gratton et al. 2007), the project also represents a response to the 2006 Equality Impact Assessment (EqIA) (Crossrail 2006). An EqIA considers whether a proposed policy or project has 'a disparate impact on persons with protected characteristics' (Pyper 2015: 22) and has been carried out by public bodies such as Crossrail, to ensure compliance with their public sector equality duty. The Crossrail EqIA focused on changes to employment, access to key services (including training and work), and disabled access. The various employment-related measures taken by Crossrail include the following:

- *Recruitment:* Encouraging contractors to recruit a more diverse workforce, by advertising all jobs externally, by working with 100 schools to encourage more young people to choose engineering as a career, by assisting in organising a 2014 National Women in Engineering Day competition, to promote engineering as a career among young women, and through a procurement policy to encourage local sourcing of goods, services, and labour;
- *Training:* Awareness training for those involved in recruitment and promotion; carrying out 'blind' recruitment; setting up a women's forum, diversity working group, and mentoring programme, and creating opportunities for senior women to act as role models for junior engineers; and assisting in organising pre-employment training at the Tunnelling and Underground Construction Academy;
- *Working and employment conditions:* Devising more inclusive maternity leave and flexible working policies, to allow staff to balance work and family life; and
- *Support:* Organising a 'women in construction' meeting in 2014, attended by 350 senior supply chain members, to share ideas about the need to support women in engineering and working with the WiC project, as on the Olympic Park.

The results have been impressive, with women representing 29% of project managers, 12% of apprentices, and 19% of graduates (Kitching 2014).

Findings: Analysis and discussion

The above three projects combined indicate the clear and strong measures an organisation needs to put in place to improve gender participation, including overarching agreements, setting of targets, and contract compliance. Figure 8.1 gives an overview of the areas involved in increasing diversity, namely employment and working conditions, recruitment, training, support measures, and DM.

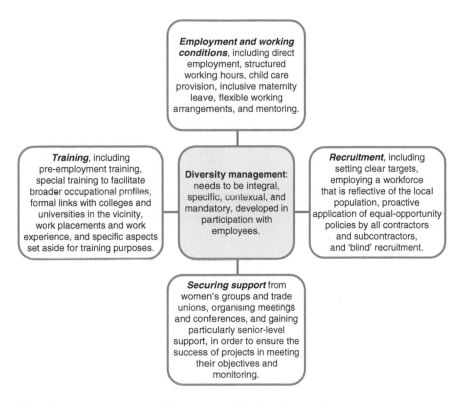

Figure 8.1 Mega projects – elements needed for effective diversity management in construction

In practice, only a minority of diversity measures is addressed ('visible'), while a significant majority is only partly addressed ('emerging'), or is not addressed at all ('hidden'). As Figure 8.2 shows, diversity measures related to 'inclusion' and formal policies are more visible than practices, recruitment-related practices are more visible than support initiatives, and gender-specific practices are more prevalent than practices linked to ethnicity or social class.

A key problem is that the structure of the construction industry itself and its current business-led DM do not provide the framework to implement the necessary measures holistically, for the 'emerging' or 'hidden' areas to become more 'visible', and therefore more effective. In order to tackle the structural determinants underlying gender segregation, research indicates that diversity strategies should focus on the umbrella of issues listed in Figure 8.2, as a holistic strategy. However, it is apparent that, while some strategies and initiatives (such as recruitment initiatives) are increasingly part of diversity practices on some prestigious mega construction projects, most strategies are not (yet) incorporated (Wright et al. 2014). Indeed, as discussed below, the structural barriers towards

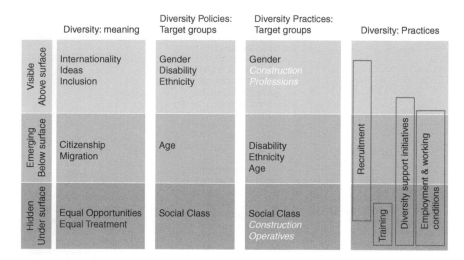

Figure 8.2 Hidden, emerging and visible measures of diversity management in construction organisations

Source: Adapted from Wright et al. (2014: 75)

more effective DM in construction are linked to the following: the lack of direct employment and employee involvement; the employer-led nature of training; the size of the company; inflexible and male-dominated employment and working conditions; and lack of support.

Employer-led DM: Who is 'the employer'?

Why is construction not conducive to gender inclusivity? One reason is that, for most of the sector, diversity policies and practices are largely an irrelevance. Indeed, there are clear indications of misplaced policies and practices put in place to increase gender participation in construction. In the first place, 91% of the 251,647 firms in the industry employed fewer than 13 female employees in 2014, and over 50% employed fewer than 3 female employees; less than 0.1% of firms employed 300 or more female employees; and only 60 firms (0.02%) employed over 1,200 female employees. Only the largest companies would be the size needed to have a human resources (HR) department with the capability to pursue an ambitious DM programme, as is also indicated in our analysis of WERS data (ONS 2016b). These larger firms, furthermore, predominantly employ only professional and technical staff, and have largely ceased to employ operatives. And almost half (924,000) of the two million-strong workforce itself come under the Construction Industry Scheme, which represents a special tax status or employment subsidy for those who are 'self-employed' rather than 'directly employed', while an unknown number of the total workforce come under agencies or are employed by labour-only subcontractors (Seely 2016). What this implies is that the employment relationship itself has been undergoing significant transformation, particularly through the use of agency labour, which includes large numbers of migrants, so that the standard employment relationship assumed to exist in DM programmes is increasingly non-standard (Janssen 2015).

The 'business case' is founded on the premise of an employer-led diversity programme, where the employer organisation is assumed to have both a hierarchical structure, to allow for CEO (chief executive officer) and senior management commitment, and an extensive HR department. As is shown by Dainty and Loosemore (2012), few construction firms are of this nature, and even those that are tend increasingly to operate as networked organisations, divided into project teams, whose members may have different employers, as is evident from our research on the Thames Tideway Tunnel (Clarke et al. 2015). Also, as is shown by Bryan et al. (2015) about large multinational construction companies, such as Multiplex and Lend Lease, institutional divisions between construction, financing and property services have become increasingly blurred, with the result that risk has been shifted onto labour, through extensive subcontracting chains. Bryan et al. (2015) argue that the finance-driven and globalised nature of the sector has involved a systematic process of risk shifting even onto the individual worker, who may well be self-employed, so that the regulatory ambiguity between 'contracts for' and 'contracts of' service is exploited; the former denotes a service or trading relationship, while the latter denotes an employment relationship (Lean 2005). In this situation, where it may be difficult for the individual worker to even identify who 'the employer' is, and where 'the employer' may be 'self-employed', the appropriateness of an employer- or business-led policy is called into question.

Diversity management: Lack of employee involvement

Without significant employee involvement to ensure that policies and practices developed are appropriate and effectively implemented, it is difficult to imagine how greater gender equality can be achieved. Ironically, the increasing onus placed on labour through subcontracting and self-employment is not matched by any increase in employee involvement in the sector, where employee representation has become weaker, with trade union members now constituting only 14% of the workforce, and with females representing only 7% of members (BIS 2015). The actions of trade unions to promote participation of women and incorporation of gender equality issues tend to be more reactive than proactive, although the Union of Construction, Allied Trades and Technicians (UCATT) in 2014 set up its first Women's Network Forum, and it also publishes a *Women in Construction* newsletter. An earlier survey of the gender equality agenda in construction, addressing the European construction social partners (employers and trade unions) across Europe, found that, while these partners have the platform to make inroads and to change the industry through collective agreements and practices that play a role in integration of women, and while they express a discourse of gender equality, this does not automatically lead to equal-opportunity policies or programmes (Clarke et al. 2005).

VET and qualifications – 'learning by doing' assumptions

The first point of entry for women into construction is through vocational education and training (VET) programmes. However, the employer-led premise of DM is mirrored in the approach to construction VET in the UK, which is also employer-based, and, following a neoclassical economic logic, focused on the demand or requirement for skilled labour, rather than on the supply side or VET provision (Keep 2015). Just as with implementation of diversity policy, this makes for a considerable disjunction between, on the one hand, the numbers of trainees to be found in reality in further education (FE) colleges and, on the other hand, forecasted skill requirements. Thus, on the supply side, the CITB's survey of first-year trainee entrants into skilled manual occupations in 2014/15 reveals the lowest intake on historical record, namely 11,586 trainees, of whom less than 5% were women, where only 35% were undertaking some work-based training, and only about 3,000 were following an apprenticeship programme (CITB 2015). This low percentage is despite estimated annual recruitment requirements for these same occupations, namely 21,790 (CITB 2016). Most VET provision, too, continues to be concentrated in the four main building trades – wood, bricklaying, painting and decorating, and plastering and drylining – although these constituted only about half of the forecasted requirements for skilled manual trades in 2015. Similar findings can be reported about professional women. For instance, 84% of the 4,830 first-year civil engineering degree students in 2012/13 were male, with just 785 female students (Wynne and Sofolarin 2014); in 2014, 71% of male graduates with engineering and technology degrees entered industry employment, while only 56% of women in the same cohort did so (EngineeringUK 2014).

Greater reliance on recruiting graduates and those – given the decline in apprenticeships – who have undertaken full-time vocational courses, means that those entering the industry depend on placements and internships to obtain work experience. But few builders, particularly the small and medium companies, take responsibility for training; 73% of construction companies have no training plan, 81% have no training budget, and only 19% invest in training (BIS 2013). As a result, recruitment into the industry should be increasingly directly from vocational colleges and universities, where higher proportions of women are found than in the labour market, but this implies much more organised and regulated work experience. That this rarely exists, in part because firms may not have the capacity, the incentive, or the resources to systematically develop and train, means that most of the few women who do undertake construction VET are lost altogether from the industry, although they may have acquired sufficient knowledge and some competencies to prepare them for work in construction.

Recruitment

Recruitment is one of the most critical HR practices that impact on changing the composition of the construction workforce, including through use of some of the proactive measures observed in our example cases, such as contract compliance, quotas, and targets. This is the reason for the strong emphasis on the 3 Rs – recruitment, retention, and respect for people – in the Rethinking Construction initiative stemming from the Egan Report of 1998 (Egan 1998; Respect for People Working Group 2002; Ness 2010). In this regard, positive action has been shown to be effective; for instance, Leicester City Council's programme for women in construction, which carefully targets women for recruitment, led to over 100 women becoming professionally qualified (EHRC 2011; Leicester City Council 2011). Nevertheless, although the informal male networks on which much recruitment has depended are weakening, informal routes to recruitment continue to predominate, particularly in SMEs, including subcontractors (Clarke and Herrmann 2007). Targeted advertising using positive female images, and more formal recruitment practices, which give greater recognition to qualifications achieved, while they are more favourable to women, are rarely found (Raymond 2013).

Employment and working conditions and work–life balance

One critical problem for women to enter and remain in the industry is the extensive use of self-employment and temporary agency working, which hampers the development of a stable workforce and clear paths of recruitment, retention, and progression (HM Treasury 2014; Harvey and Behling 2008). Much of the success in improving diversity on projects such as the Olympic Park and T5 was attributable to the insistence on direct employment. Another problem for women is the long-hours culture and expectation of total availability (Watts 2009; Wright 2014b); as Ness (2012: 668) argues, 'the exclusion of women both enables and condemns men to work long hours'. Long-hours working is therefore not only an obstacle to women's participation, but is also the product of their historical exclusion from the world of work (GLA 2007). People in the industry work many hours more than their contract states they should, where presenteeism is still a core part of the working culture, and long-hours working is seen as an indicator of commitment, and is therefore used as a prerequisite for promotion (Wright 2014b). On average, a construction worker, including those that work part-time, puts in 41.2 hours a week, compared to an average of 36.3 hours across the UK economy, with crane drivers having the longest hours of any single occupation, working on average 52.8 hours a week (Gardiner 2012). These long hours go together with higher earnings, so that average total weekly earnings in construction (£581) are also higher than the average for the whole economy (£528) (ONS 2016b).

The UK science and engineering sectors have their female workforce exit the sector at a much higher rate than other sectors. A major contributing factor to this loss is the lack of part-time work opportunities, with only 12% of female engineering professionals working part-time in 2010, compared with 42% of all UK female employees (Hart and Roberts 2011). The lack of flexible working as an opportunity to balance work and caring roles is a key element in explaining female retention problems (Ryan and Kossek 2008). The masculine nature of the work environment, particularly on-site work, where humour and sexualised banter in the workplace is common, can also prevent female inclusion (Faulkner 2009). Other concerns of diversity policies are the positioning of women as role models and the putting in place of successful mentoring schemes, so helping to convey the organisation as an equal-opportunities employer and demonstrating that perceived barriers to progression are not insurmountable. However, given the low numbers of women in many construction occupations, all such policies appear as tweaking, rather than as seeking to transform an industry structured to exclude women into one that is gender-neutral. Although where sectoral and organisational policies for support are made available, progress is visible (English and Hay 2015).

Summary

What we have sought to show is that the employer-centred approach, on which the 'business case' for greater gender participation through diversity policies and practices is founded, is inappropriate for the construction sector. Not only is it increasingly difficult to identify who 'the employer' is, but the current structure of the industry is such that diversity policies and practices cannot be adequately implemented. Self-employment, extensive subcontracting, networked organisations, and long-hours working are simply at odds with the traditional hierarchical organisation, with its wide-ranging HR department. Achieving gender equity is only possible through a transformation of the industry, not just through reform of existing employment norms, but also by expanding its priorities and objectives, from being concerned primarily with profit to encompassing a wider set of ethical considerations, including stable employment, training and apprenticeships, and sustainable construction. Forces for change include the need for greater educational input and broader qualification profiles, if the more abstract competencies required for today's construction labour process are to be developed. Bridging the interfaces between the activities of different professionals and occupations and integrated team-working and communication are also more and more essential, given the complex work processes involved. This implies a radical transformation of the construction process, affecting all occupations, and opening up the possibility to include more women, particularly considering their higher educational achievements and greater presence in technical and environmental subject courses.

References

BAA *see* British Airports Authority.
Bagilhole, B. (1997) *Equal opportunities and social policy: Issues of gender, race and disability.* London: Longman.

Baker Mallett (2008) *BAA Heathrow Terminal 5 Programme: The Major Projects Agreement.* End of Programme Audit Report, March.

Barnard, S., Powell, A., Bagilhole, B. and Dainty, A. (2010) Researching UK women professionals in SET: A critical review of current approaches. *International Journal of Gender, Science and Technology* 2(3).

BIS (2013) *UK construction: An economic analysis for the sector.* Department for Business, Innovation and Skills, July.

BIS (2015) *Trade union membership 2014.* Department for Business Innovation and Skills, Statistical Bulletin, June.

Briscoe, G. (2005) Women and minority groups in UK construction: Recent trends. *Construction Management and Economics* 23(10): 1001–1005.

British Airports Authority (BAA) (2004) *T5 Agreement: We're making history.* T5 Internal Communication Team, April.

Bryan, D., Rafferty, M., Toner, P. and Wright, S. (2015) Risking it all – Financialisation and labour in the construction industry. *CLR News* 1: 12–22.

CIPD (2006) *Diversity in business: A focus for progress.* Survey report. March. London: CIPD.

CIPD (2007) *Diversity in business: A focus for progress.* Survey report. March. London: CIPD.

CITB (2015) *Training and the built environment 2015.* Bircham Newton: CITB.

CITB (2016) *Industry insights: Construction skills network forecasts 2016-2020.* Bircham Newton: CITB.

Clarke, L. (2007) The emergence and reinforcement of class and gender divisions in vocational education, in Clarke, L. and Winch, C., *Vocational education: International approaches, developments and systems.* Abingdon, UK: Routledge, pp. 62–76.

Clarke, L. and Gribling, M. (2008) Obstacles to diversity in construction: the example of Heathrow Terminal 5. *Construction Management and Economics* 26(10): 1055–1065.

Clarke, L. and Herrmann, G. (2007) Skill shortages, recruitment and retention in the housebuilding sector. *Personnel Review* 36(4): 509–527.

Clarke, L. and Holborough, A. (2011) The forthcoming Olympics in London. *CLR News* 2: 44–53.

Clarke, L. and Wall, C. (2009) 'A woman's place is where she wants to work': Barriers to the entry and retention of women into the skilled building trades. *Scottish Labour History* 44: 16–39.

Clarke, L. and Wall, C. (2011) Skilled versus qualified labour: The exclusion of women from the construction industry, in Davis, M. (ed.), *Class and gender in British labour history.* London: Merlin Press.

Clarke, L. and Wall, C. (2014) Are women 'not up to' working in construction – at all times and everywhere?, in Munn, M. (ed.), *Building the future: Women in construction.* London: The Smith Institute. pp. 10–19.

Clarke, L., Frydendal Pedersen, E., Michielsens, E. and Susman, B. (2005) The European construction social partners: Gender equality in theory and practice. *European Journal of Industrial Relations* 11(2): 151–177.

Clarke, L., Frydendal Pedersen, E., Michielsens, E., Susman, B. and Wall, C. (eds) (2004) *Women in construction.* Brussels: Reed Business Information.

Clarke, L., Michielsens, E. and Wall, C. (2006) Women in manual trades, in Gale, A. and Davidson, M. (eds), *Managing diversity and equality in construction: Initiatives and practice,* pp. 151–168. London: Routledge.

Clarke, L. Michielsens, E., Snijders, S., Wall, C. (2015) 'No more softly, softly': Review of women in the construction workforce. London: Centre for the Study of the Production of the Built Environment (ProBE), University of Westminster.

Cox, T. and Blake, S. (1991) Managing cultural diversity: Implications for organizational competitiveness. *The Executive* 5(3): 45–56.

Craig, S. and Oates, A. (2014) Empowering women in construction, in Munn, M. (ed.), *Building the future: Women in construction*. London: The Smith Institute. pp. 78–86.

Crossrail (2006) *Crossrail Equality Impact Assessment: Project & Policy Assessment Report: Briefing*. Department for Transport, January.

Dainty, A. and Loosemore, M. (eds) (2012) *Human resource management in construction: Critical perspectives*. Abingdon, UK: Routledge.

Dainty, A.R.J., Bagilhole, B.M. and Neale, R.H. (2000) A grounded theory of women's career under-achievement in large UK construction companies. *Construction Management and Economics* 18(2): 239–250.

Dickens, L. (1999) Beyond the business case: A three-pronged approach to equality action. *Human Resource Management Journal* 9(1): 9–19.

Drake, B. (1984) *Women in trade unions*. London: Virago.

Druker, J. and White, G. (2013) Employment relations on major construction projects: The London 2012 Olympic construction site. *Industrial Relations Journal* 44(5–6): 566–583.

Egan, J. (1998) *Rethinking construction*. Construction Task Force Report for the Department of the Environment, Transport and the Regions. London: HMSO.

EHRC *see* Equality and Human Rights Commission.

EngineeringUK (2014) *Engineering UK 2014: Synopsis, recommendations and calls for collaborative action*. London: EngineeringUK.

English, J. and Hay, P. (2015) Black South African women in construction: Cues for success. *Journal of Engineering, Design and Technology* 13(1): 144–164.

English, J. and Le Jeune, K. (2012) Do professional women and tradeswomen in the South African construction industry share common employment barriers despite progressive government legislation? *Journal of Professional Issues in Engineering Education and Practice* 138(2): 145–152.

Equality and Human Rights Commission (EHRC) (2011) *Equality and diversity: Good practice for the construction sector*. Report by Jan Peters, May.

Experian (2006) *Employment and skills for the 2012 Games: Research and evidence*. Final Report to the Learning and Skills Council and the London Development Agency, May.

Faulkner, W. (2009) Doing gender in engineering workplace cultures. I. Observations from the field. *Engineering Studies* 1(1): 3–18. doi: 10.1080/19378620902721322

Fielden, S.L., Davidson, M.J., Gale, A.W. and Davey, C.L. (2000) Women in construction: The untapped resource. *Construction Management and Economics* 18(1): 113–121.

Foster, C. (2010) The ODA: Increasing number of women working in construction. *Equal Opportunities Review* September, No. 204, pp. 8–10.

Francis, V. (2015) What helps professional women advance in construction? *CLR News* 3: 21–24.

Gardiner, J. (2012) Should we work all hours? *Building*, 20 July.

GLA *see* Greater London Authority.

Gratton, L., Kelan, E., Voigt, A., Walker, L. and Wolfram, H-J. (2007) *Innovative potential: Men and women in teams*. London: The Lehman Brothers Centre for Women in Business, London Business School.

Greater London Authority (GLA) (2007) *The construction industry in London and diversity performance*. February, GLA Mayor of London.

Hart, R. and Roberts, E. (2011) *British women in science and engineering: The problem of employment loss rates*. Working paper. Stirling, UK: Division of Economics, University of Stirling.

Harvey, M. and Behling, F. (2008) *The evasion economy: False self-employment in the UK construction industry*. London: UCATT.

HM Treasury (2014) *Self-employment in the construction industry*. Business and Transport Section.

Hoque, K. and Noon, M. (2004) Equal opportunities policy and practice in Britain: Evaluating the 'empty shell' hypothesis. *Work, Employment and Society* 18(3): 481–506.

Janssen, J. (2015) Instead of an introduction: The transformation of employment in construction. *CLR News* 1: 7–11.

Kandola, A. and Fullerton, J. (1998) *Managing the mosaic: Diversity in action*. Various editions. London: CIPD.

Keep, E. (2015) Governance in English VET: On the functioning of a fractured 'system'. *Research in Comparative and International Education* 10(4): 464–475.

Kitching, R. (2014) Women in engineering: Leading the charge, in *New Civil Engineer* 1 July.

Kochan, T., Bezrukova, K., Ely, R., Jackson, S., Joshi, A., John, K., Leonard, J., Levine, D. and Thomas, D. (2003) The effects of diversity on business performance: Report of the diversity research network. *Human Resource Management* 42(1): 3–21.

Lean, G. (2005) When does a contract for services become a contract of employment? *CLR News* 2: 47–52.

Leicester City Council (2011) Women in construction courses. Available at: www.leicester.gov.uk

Lingard, H. and Francis, V. (2004) The work-life experiences of office and site-based employees in the Australian construction industry. *Construction Management and Economics* 22(9): 991–1002.

London 2012 (2011) *Women into construction*. London 2012.

Michielsens, E., Wall, C. and Clarke, L. (1997) *A fair day's work: Women in the direct labour organisations*. London: Women and the Manual Trades.

Ness, K. (2010) The discourse of 'Respect for People' in UK construction. *Construction Management and Economics* 28(5): 481–493.

Ness, K. (2012) Constructing masculinity in the building trades: 'Most jobs in the construction industry can be done by women'. *Gender, Work & Organization* 19(6): 654–676.

ODA *see* Olympic Delivery Authority.

Office for National Statistics (2014) *EMP13: Employment by industry*. 13 August.

Office for National Statistics (2016a) *Construction statistics: No 17, 2016 edition*. London: ONS.

Office for National Statistics (2016b) *Annual survey of hours and earnings: 2016*. 20 July.

Olympic Delivery Authority (ODA) (2011) *Jobs, skills, futures – Employment and skills update*. London: ODA.

ONS *see* Office for National Statistics.

Özbilgin, M. and Tatli, A. (2011) Mapping out the field of equality and diversity: Rise of individualism and voluntarism. *Human Relations* 64(9): 1229–1253.

Payne, J. (1991) *Women, training and the skills shortage: The case for public investment*. London: Policy Studies Institute.

Pyper, D. (2015) *The public sector equality duty and equality impact assessments*. House of Commons Library. Briefing Paper Number 06591. 22 January.

Raymond, J. (2013) Sexist attitudes: Most of us are biased. *Nature* 495(7439): 33–34.

Respect for People Working Group (2002) *A framework for action*. Rethinking Construction Report Housing Forum.

Ross, R. and Schneider, R. (1992) *From equality to diversity: A business case for equal opportunities*. London: Pitman.

Ryan, A.M. and Kossek, E.E. (2008) Work-life policy implementation: Breaking down or creating barriers to inclusiveness? *Human Resource Management* 47(2): 295–310.

Sang, K.J.C. and Powell, A. (2012) Equality, diversity, inclusion and work-life balance in construction, in Dainty, A. and Loosemore, M. (eds), *Human resource management in construction: Critical perspectives*. Abingdon, UK: Routledge. pp. 163–196.

Seely, A. (2016) *Self-employment in the construction industry*. House of Commons Library. Briefing Paper No. 000196. 23 May. London.

The Construction Index (2012) Women still under-represented in construction. *The Construction Index* 16 May.

UKRC (2005) *Guidance booklet for teaching and professional staff in HE and FE: encouraging gender equality in STEM*. Bradford, UK: UK Resource Centre for Women in Science, Engineering and Technology.

Wall, C. and Clarke, L. (1996) *Staying power: Women in direct labour building teams*. London: Women and the Manual Trades.

WAMT (2001) *Building the future: Twenty-five years of Women and Manual Trades*. London: WAMT.

Watts, J.H. (2009) 'Allowed into a man's world' meanings of work–life balance: Perspectives of women civil engineers as 'minority' workers in construction. *Gender, Work & Organization* 16(1): 37–57.

Wright, A., Clarke, L., Michielsens, E., Snijders, S., Urwin, P. and Williamson, M. (2014) *Diversity in STEMM: Establishing a business case*. University of Westminster Report for Royal Society Diversity Programme, June 2014.

Wright, T. (2013) Promoting employment equality through public procurement. Report of a workshop held by the Centre for Research in Equality and Diversity (CRED), Queen Mary, University of London.

Wright, T. (2014a) *The Women into Construction Project: An assessment of a model for increasing women's participation in construction*. Centre for Research in Equality and Diversity, Queen Mary, University of London.

Wright, T. (2014b) Gender, sexuality and male-dominated work: The intersection of long-hours working and domestic life. *Work, Employment and Society* 28(6): 985–1002.

Wynne, A. and Sofolarin, Y. (2014) Engineering equality: Are universities learning the gender lesson? *New Civil Engineer* 19 August.

9 Women in the construction industry

Still the Outsiders

Erica French and Glenda Strachan

Space and time are the framework within which the mind is constrained to construct its experience of reality

(Immanuel Kant)

Background

The construction industry is characterised by extreme gender segregation, both horizontal and vertical. Internationally, women represent a small proportion of employees in construction, for example, 8.9 per cent of workers in the USA, 11.5 per cent in Canada, and 14.2 per cent in Japan (Catalyst, 2015). Evidence shows that not only are proportions small now, the change in figures over time is also limited. In 2002–03 women comprised 9 per cent of the construction workforce in the UK but within this there is significant horizontal segregation by occupation, with 84 per cent of these employees in secretarial work and 10 per cent in professional categories such as design and management (Gurjao nd, 16). In 2015 Randstad (2016) reported the number of women in construction in the UK as 20 per cent, with the numbers of women in management in construction moving from 6 per cent in 2005 to 16 per cent in 2015. But this report, compiled by a company within the industry, is not supported by the UK construction union, the Union of Construction, Allied Trades and Technicians (UCATT) who note that women form only 11 per cent of the industry with only 1 per cent working on a construction site in 2015. In the USA figures remain unchanged over the past five years with the United States Department of Labor identifying 9 per cent of women in construction in 2010. In Australia in 2016 11.7 per cent of employees in the industry were women (WGEA, 2016a). The proportion of women in the industry had decreased since 1995 when women formed 14.8 percent of the construction workforce (WGEA, 2016b). It is increasingly clear that the percentage of women in construction is low internationally and little has changed in the past few decades.

The sparse literature which deals with the construction industry discusses two threads about gender – getting women into the industry and keeping them there.

Once women enter the industry, particularly in typically male occupations including management, the literature discusses the issues women face. The primary reason for the lack of women in the industry identified in the literature, predominantly from the UK, is the image and reputation of a male culture (Cartwright & Gale, 1995; Dainty et al., 2001; Fielden et al., 2001; Worral et al., 2010). These issues are similar to those experienced by women in other male-dominated industries and occupations but are often heightened by the extreme nature of the segregation.

Predominantly qualitative studies have revealed that the male-dominated culture and inflexible work practices are major issues (Worrall et al., 2010). In the UK interviews conducted with women and men uncovered human resource management (HRM) practices which maintain current workplace environments with men resisting changes to the construction culture which has supported them (Dainty et al., 2001), particularly through improved chances of promotion and rapid career development due to opportunities for work overseas. But it is not only men who resist that culture change. In a study interviewing female engineers, Watts (2009) identifies two themes that are central to the conflict many women face in construction, namely the long hours culture, and managing the work and family divide. Moreover, Watt (2009) acknowledges that because women are adopting similar work styles to those of their male colleagues, they may not be the agents of change in this industry. The poor image and reputation of the construction sector is the primary barrier, according to Fielden et al. (2001), and word-of-mouth recruitment, limited terms and conditions of employment, lack of training and male networks presented further hurdles for women.

Unsurprisingly, there are very few women in senior management in the construction industry. Using the findings from 2,000 surveys of female and male surveyors, Ellison (2001) found that, despite equal education qualifications, women remain under-promoted in comparison with men, even though women invest time, money and effort into the advancement of their careers. Further, in a study of project managers including both men and women in engineering, construction and information technology, Crawford, French, and Lloyd-Walker (2013) found women experience their careers in these industries differently. Not only was this the perception of women but also that of men where "…two thirds of respondents believe women experience difficulties in their project manage-ment careers related to lower pay; fewer opportunities; and less support". Both men and women equally discussed the importance of their individual career development and their commitment to the 'profession' and their careers.

In their examination of engineering and construction sectors in the UK, Powell and Sang (2015) revealed the gendered treatment that women experi-ence in everyday interactions at work. They conclude that "…women's differ-ence from men is reiterated and experienced as a matter of routine… [and this] has rendered this sexism largely invisible for younger women" (Powell & Sang, 2015: 931). While English and Hay (2015) found some positive measures for women in the construction industry in South Africa, "…two out of three respondents felt that construction organisations are structurally and culturally male, where long working hours and norms support a workplace culture of inflexibility and discrimination" (English & Hay, 2015: 156). Women who do remain in the industry often occupy particular niches to avoid the male culture (Gale, 1994) or develop bespoke long-term careers for individuality to avoid any resistance through the male-dominated culture (Dainty et al., 2001).

Some writers are now calling for organisational policy, practice and culture change. English and Hay (2015: 160) conclude that "...political policy is insufficient to elicit change; monitored equal opportunity actions that target women are required". Others argue that policies and practices which focus on 'fixing the women' so that they fit in with the extant industry and organisational practices have shown little benefit. Powell and Sang (2015) question existing policy recommendations which argue that women bring different skills to the sector and those which focus on increasing numerical diversity. They argue that these policies reinforce the gendered nature of the industry and "...fail to recognise how the underlying structures and practices of the sector reproduce gendered working practices" (Powell & Sang, 2015: 932). From an analysis of property industry annual reports in Australia, Warren and Antoniades (2015) conclude that attitudinal and structural change is needed to move towards gender equality.

In this chapter, we seek to identify what construction organisations are doing to manage equality and inclusion in a highly male-dominated industry. We use national reporting data from Australia. Australia can be utilized as a country case study of organisational interventions in the construction sector as it has had national legislation since 1986 which mandates reporting on organisational programmes to advance gender equity. This study investigates how three legislative changes have impacted on implementation approaches to equity management in Australia. It specifically focuses on the extent of equity policies in construction companies and asks the question whether there have been any changes for women in these organisations.

Equity legislation in Australia

There have been three Acts of Parliament which have focused on gender equity in employment since 1986, but the objectives have been virtually the same and all have focused on mandatory reporting of organisational policies. The three Acts promoting gender equity come in addition to anti-discrimination legislation. This legislation was seen by the government as insufficient to achieve gender equality and the rationale for the legislation relied on the disadvantaged position of women in the workforce and utilised the concept of systemic discrimination (Department of Prime Minister and Cabinet, 1984: vol. 1, 12–13). The *Affirmative Action (Equal Opportunity for Women) Act 1986* was "...about achieving equal employment opportunity for women. To achieve this goal, the barriers in the workplace which restrict employment and promotion opportunities for women have to be systematically eliminated" (South Australian College of

Advanced Education and Affirmative Action Agency (Australia), 1990: 1). This section provides a brief outline of these Acts and their reporting focus.

The three Acts are characterised by similarities in goals although the names of the Acts differ. All are based on the concept of merit or the 'best person for the job' and do not contain quotas. All require organisational reporting from organisations with more than 100 employees on an annual basis (with some two-yearly reporting under some Acts for those with recognised good performance). To achieve the aims of advancing gender equity, the legislation relies on organisational goodwill to devise and implement policies. While reporting is mandatory, there were no mandatory policies until the 2012 Act, outside of those which respond to other pieces of legislation, such as industrial law and anti-discrimination legislation. There have been weak compliance mechanisms, and these operate essentially when a company does not submit a report or, under the 2012 Act, it fails to meet minimum standards for two consecutive years. The consequences of non-compliance are naming in Parliament and preclusion from carrying out business with the federal government.

What has differed from Act to Act is the way that organisations are asked to report. Organisations have always been required to report the number of employees by employment category and gender. However, the types of employment in the reporting forms have differed somewhat under each Act, thus making nuanced comparisons difficult. The requirements and methods for reporting organisational policies have also changed. The 1986 Act specified that organisations undertake an eight-step programme which included an analysis of the position of women in their organisation through examination of employment statistics, personnel practices – both written and unwritten – and consultation with women employees and trade unions. Based on this analysis, each company was required to devise a programme which addressed some of the problems identified and set targets against which future progress could be judged (Strachan, 1987). Reporting required organisations to comment on what they had done in each of these eight steps and included a list of HR policies such as recruitment and flexible leave arrangements which the organisation could tick if they had them (Strachan & Burgess, 1999). French (2001), in an extensive study of all organisations who submitted a report to the Affirmative Action Agency in 1997 (N=1976), found anti-discrimination strategies, including equal treatment of women and men through gender-blind strategies and policies, were not significant predictors of increases in women's level of employment in management. Affirmative action (equal opportunity) strategies, including different treatment and gender conscious strategies, were predictors of significant increases in management levels for women. Less than 22 per cent of organisations at that time used such approaches in their policies and strategies in HR structures, work organisation or conditions of service, and in few areas of influence. However, as a large combined study, it precluded the identification of industry differences in implementing equal employment opportunity and, in the absence of longitudinal measurement, it limited the causal inferences and application of the findings.

The *Equal Opportunity for Women in the Workplace Act 1999* replaced the 1986 Act but retained similar goals. An equal opportunity for women in the workplace programme was interpreted as appropriate action to eliminate discrimination against women and employer-initiated measures to contribute to the achievement of equal opportunity for women (Section 3). However, the 1999 Act removed the eight steps outlined in the previous Act as well as the consultation with trade unions and dispensed with specific goals or targets in favour of a more general, "user-friendly" approach (Strachan et al., 2004). Organisations were required to develop a workplace programme by preparing a workplace profile, analysing the equity issues for women, identifying priority issues, taking action to address them and evaluating the effectiveness of the actions. They had to address seven employment matters: recruitment and selection; promotion, transfer and termination; training and development; work organisation; conditions of service; arrangements for dealing with sex-based harassment; and arrangements for dealing with pregnancy, potential pregnancy, and breastfeeding. Reporting required an organisation to discuss what they were doing under each of these employment issues, and not just tick a list of activities. French and Strachan (2007, 2009 and 2015) report on three industry studies in the finance, transport, and construction industries. Analysis showed that industry differences in the implementation of equal opportunity measures do occur. Male-dominated industries of transport and construction use equal treatment strategies for equal opportunity implementation, and these are not linked to significant increases of women in management or non-traditional areas of work. However, finance organisations, typically a female-dominated industry, did utilise more special measures in approaches to equal opportunity but these occurred in work organisation (flexible work) rather than in the HR areas of recruitment or promotion. No significant increases were noted of women in management or non-traditional areas of work, but more women were able to move in and out of organisations to undertake caring roles.

The *Workplace Gender Equality Act 2012* uses the term 'gender' rather than 'women', thus moving from women only to include men. Its first two objectives are to "...promote and improve gender equality (including equal remuneration between women and men) in employment and in the workplace" and to "... support employers to remove barriers to the full and equal participation of women in the workforce, in recognition of the disadvantaged position of women in relation to employment matters". The 2012 Act aims to focus on outcomes, what an organisation is doing, while the previous Acts focused on the process of organisational plans. The 2012 Act sets out five gender equity indicators (GEIs): gender composition of the workforce; gender composition of the governing body; equal remuneration between women and men; availability and utility of working arrangements which are flexible or support employees with family or caring responsibilities; and consultation with employees on issues concerning gender and equality at the workplace. It retains the "light-handed" model of compliance (Sutherland, 2013).

Under each GEI, minimum standards are created which represent the minimum that organisations with more than 500 employees must do to demonstrate their commitment to gender equality. In practice, this has translated into a reporting instrument that organisations tick. For instance, "Do you have formal policies or formal strategies in place that specifically support gender equality about" recruitment, retention, and so on (WGEA, 2016c). While there is a space on the form for more explanation, in practice this is utilised infrequently by construction firms. Therefore, the current forms provide little or no detail about what policies and strategies are being utilised, and make no mention of what the priorities are for change in the organisation.

Methodology

This research involves two studies, each using secondary data. The first study analysed information provided by all construction organisations (n=90) reporting in one year to the Australian Government under the *Equal Opportunity for Women in the Workplace Act 1999* on their equity management practices. These were analysed through content analysis using the typology described in Table 9.1. In 2011–12, 90 construction organisations submitted equal employment opportunity (EEO) progress reports to the Agency. Errors and omissions in seven left us with 83 viable reports. Each progress report becomes a public document and must detail the workplace profile of men and women and their job roles, the equal employment issues specific across seven employment matters (named below), and the organisational policies and strategies for addressing these issues as well as priorities of actions taken and plans. For this study, appropriate reports were downloaded from the Agency's Online Searchable Database of Reports in May 2012 (available at WGEA, 2016e).

Measures:

- **Employment profile.** Employment details of men and women in specific job roles were aggregated to four main categories: management (including senior executives, management, supervisory staff and professional staff), operations (including maintenance, technicians, trades and miscellaneous personnel), sales, and clerical staff.
- **Equality and diversity approaches.** The seven employment matters reported on were recruitment and selection, promotion and transfer, training and development, work organisation, conditions of employment, addressing sexual harassment, pregnancy and breastfeeding policies. Information on each of the seven employment matters was classified according to the equal opportunity approach taken by the organisation.
- **Organisational size.** Organisation size has been considered to be a significant predictor of the employment status of women (Konrad & Linnehan, 1995; French, 2001). Using four categories used by the Agency, we measured size as the number of employees ranging from 100–500, 500–1000, 1000–3000, and 3000 or more, and took the natural logarithm of the midpoint of each category for use in the analyses.

Table 9.1a Typology of organisational policies designed to promote gender equity. Based on French (2001)

Goals	Instruments
Type 1 – No reporting: No comments on employment equity, equal treatment or equal results and no recognition of individual difference or disparity. No issues identified or no strategies outlined at all on any employment matters.	No policy instruments. The organisation supports the current situation, with or without acknowledgement of any discrimination or disadvantage in that situation.
Type 2 – Traditional approach: The traditional (or classical) classification refutes discrimination plays any role in workplace disparity between different employees (or groups) and supports the different treatment of individuals in the workplace based upon their individual choices. This approach advocates against the specific implementation of equity measures, instead calling on women and minority groups to make different educational and lifestyle choices in order to create change (French, 2001).	No policy instruments, merely an objective to refute discrimination as a contributory factor in workplace disparity. Acknowledgement of individual difference in choices. In this study comments such as, *'Recruitment and selection is always based on the best match between the prospective candidate to the skills and competencies set out in the job description'*; *'Women are mainly employed in clerical positions'*; and *'When vacancies arise they are advertised externally and internally to ascertain the best person for the position'*.
Type 3 – Anti-discrimination approach: The anti-discrimination classification acknowledges the importance of the removal of discriminatory practices and processes in order to offer equal treatment based on human rights principles. This approach fulfils the requirements of anti-discrimination legislation such as the Sex Discrimination Act 1984. Activity is limited to equal treatment and/or equal outcomes for men and women (French, 2001; Konrad & Linnehan, 1995).	Equal treatment policies and practices evidenced across some or all the human management practices fulfilling a strategy of equal treatment. In this study comments such as *'No [job] advertisement is gender biased'*; *'All staff have attended seminars on harassment and are aware of the responsibilities and their rights under the policy'*; and *'7 of the 9 women on maternity leave have returned to work either in their previous position or a part time position for an agreed period of time'*. Also reported comments such as *'Our policy is to treat men and women equally'* were included into this category.

(Continued)

Table 9.1a Typology of organisational policies designed to promote gender equity. Based on French (2001) (Continued)

Goals	Instruments
Type 4 – Equal employment opportunity (EEO) approach: The EEO classification acknowledges the importance of the removal of discriminatory practices as well as the adoption of special measures which are identity conscious and designed to assist members of disadvantaged groups – in this case women. This follows the usage of the term 'affirmative action' based on recognition and acceptance of the fact that it is not sufficient to make specific acts of discrimination unlawful. "Further steps are needed to relieve the effects of past discrimination, to eliminate present discrimination and to ensure that future discrimination does not occur" (Department of the Prime Minister and Cabinet, 1984, p. 8). Supports Konrad and Linnehan's (1995) findings which identified gender conscious treatments as having different outcomes for women in organisations.	Identity conscious policies and practices (or different treatment) of specific groups based on differences in outcomes that may be historical, organisational or social in nature as well as identity blind (equal treatment) to address potential discrimination fulfilling a different treatment strategy for different groups. In this study reports on specific strategies such as apprentice or graduate programmes for the recruitment of women were classified as EEO in nature. Examples include: *'We attempt to ensure that there is a female employee on the interviewing panel to ensure that all applicants are given a fair go'*; and *'A mentoring process has been established, including coaching with study and career guidance and advice for a number of female employees'*.
Type 5 – Gender diversity approach: The gender diversity classification acknowledges the potential for bias and discrimination against women within organisational structures and supports the neutral treatment of all individuals based on organisational requirements as a means of addressing any discrimination. While there is debate about exactly what constitutes policies and programs variously labelled 'diversity' and 'managing diversity' (Bacchi, 2000; Kirton & Greene, 2005), we have used the term 'gender diversity' to incorporate elements of organisational/structural change. In order to classify policies as gender diversity, organisations needed to include elements of culture change within the organisation.	Diverse policies and practices based on organisational requirements and differences between employees ensuring neutral treatment of groups and of individuals, encouraging greater flexibility and inclusivity with equal access to benefits and burdens of organisations and addressing culture change. In this study reports on policies and practices, such as leave opportunities that were the same for both genders, were classified as diverse in nature. Examples included: *'The processes established for consideration of individual needs in relation to work organisation and rostering have operated effectively this year with management, and unions combining efforts to ensure that problems and grievances were effectively resolved'*; *'Workplace flexibility is considered by balancing employee needs particularly those related to family with the organisation needs'*; *'Every effort is made to provide employees with a means to balance work and family responsibilities including providing job sharing, flexible working hours, carer's leave and recognising the need to minimise overtime'*; and *'We continue to provide remote access to the company's computer systems so that staff with family responsibilities can work from home'*.

To determine any relationship between the dependent variable (DV) and the independent variables (IVs), ordinary-least-square (OLS) regression analysis was used. IVs consisted of the approach taken, action was taken, priority and future actions. The DVs consisted of the specific numbers of women and men in specific job roles. Regression analyses reveal relationships among variables without implying causality. In this case, the regression analysis identifies the relationship (if any) between policy implementation and the position of women and men in construction organisations and allows the prediction of such occurrences.

The second study involved the 20 major construction firms in Australia, based on the list of the Top 20 Construction companies published by the Australian Mines and Metals Association (AMMA), which is Australia's largest national resource industry employer group (AMMA, 2016). For this study, the appropriate reports were downloaded from the Workplace Gender Equality Agency's Online Searchable Database of Reports (WGEA, 2016e). We analysed each report under the same typology along the six reportable gender equality indicators which under the new legislation include: gender composition of the workforce; gender composition of the governing bodies; pay gap; availability and utility of employment terms, conditions and practices relating to flexible working arrangements supporting employees with family or caring responsibilities; consultation with employees on issues concerning gender equality in the workplace; and sex-based harassment and discrimination interventions.

Public availability of the reports under both iterations of the legislation and the potential to be named in Parliament for a non-compliant report may be seen as pressure to present a socially desirable image through individual reports. Social desirability bias (Fowler, 1988) is a recognised threat to an accuracy of information when there is pressure to present a socially desirable image of organisations. The legislation (the Equal Opportunity for Women in the Workplace [EOWW] Act 1999 and the Workplace Gender Equality Act [WGEA] 2012) attempts to ensure accuracy of information through the mandatory requirement of the signatures of both the report writer (usually the HR manager) and the CEO on all reports submitted to the Agency. Further, each report is checked at the Agency, evaluated and the organisation contacted to verify information, make recommendations and give feedback. Trained assessors review report contents to first ensure compliance status under the Act, then evaluate the organisation's analysis of equal employment issues and identify demonstrated links with the organisation's current actions and plans. Information is provided for those organisations not compliant under the Act to assist them to meet compliance standards (EOWA, 2006).

Quantitative analysis of qualitative data can potentially prove a threat to accuracy and reliability as there is the possibility that the researcher may 'force' cases into categories that reflect the biased views of the researcher rather than the substantive actions of the respondents (Crompton & Harris, 1999). To address this issue we used the pre-determined typology of approaches to determine the categories into which the responses were to be divided and generated an appropriate coding scheme on this basis (Harris, 2001). To address reliability, the coding process was separated from the process of data entry to allow for cross-checking. Besides, both researchers worked together on the coding process, with one researcher checking a sample of the coding from the other (Krippendorff, 1980).

Results and discussion

Under the 1999 Act, there was evidence of a range of different policies and practices adopted by organisations in construction to address the seven employment matters reportable (see Table 9.2). In the areas of 'recruitment and selection', 'promotion and transfer' and 'training and development', 42.7 per cent, 74.4 per cent and 65.4 per cent of organisations respectively reported they had no activities of any type to address any perceived disparity or inequality of women in their organisations. Combined with the percentage of organisations that refused to report in these areas, the result indicates more than half of all construction organisations had a less than compliant level of activities in the areas of recruitment and selection, training and development, and promotion and transfer to address discrimination or bias against women. Limited numbers of organisations (13.4 per cent, 11.0 per cent and 12.3 per cent respectively) took an equal treatment approach to these employment matters which addresses discrimination and encourages equal treatment and access to opportunity. This involved the use of practices that encouraged the equal treatment of men and women in recruitment, promotion and development, such as women on selection panels and equal numbers of men and women offered access to development opportunities. A limited number of organisations took a proactive approach of some kind – either special consideration activities specifically for women, or neutral treatment programmes embedded in organisational flexibility and change in designing and delivering opportunity strategies specific to the disadvantage of either or both men and women. In the areas of 'work organisation' and 'conditions of service' some organisations offered no specific strategies for addressing inequity. In those organisations that did seek to proactively address inequality through EEO, special measures or gender diversity strategies, approximately 70 per cent took action to develop equitable work patterns (work flexibility), while only 35 per cent took action to develop fairness in conditions of service (pay and conditions) (see Table 9.2).

In the area of addressing harassment, the majority of organisations took a compliance-based approach in ensuring equal treatment through training of all staff, regardless of gender or organisational role. While some were not compliant, these were in the minority. This is not surprising given the strength of provisions in the *Sex Discrimination Act 1984*, which defines and prohibits discrimination and harassment by gender and outlines extensive provisions for obtaining justice. Further, the tribunals and courts emphasise the importance of appropriate policies and practices and are supporting zero tolerance through judgments awarding

Table 9.2 Organisations' approaches to equality in diversity by percentage

Approach type	R&S	Promote transfer	T&D	Work organisation	Conditions of service	Sexual harassment	Pregnancy & breastfeeding
Nil – no strategies	1.2%	3.7%	1.2%	1.2%	1.2%	1.2%	1.2%
Traditional – no instruments	42.7%	74.4%	65.4%	29.3%	64.6%	3.7%	7.3%
Anti-discrimination	13.4%	11%	12.3%	0%	17.1%	89.0%	25.6%
Equal treatment universal							
EEO – equal treatment universally and special consideration policies for different groups	18.3%	1.2%	7.4%	12.2%	1.2%	2.4%	11.0%
Gender diversity neutral treatment through equal access to increasing flexibility options	0%	0%	0%	52.4%	12.2%	1.2%	53.7%
Combination strategies without special consideration	1.2%	3.7%	1.2%	3.7%	1.2%	0%	0%
Combination with special consideration	23.2%	6.1%	12.3%	1.2%	2.4%	2.4%	1.2%

R&S = Recruitment and selection; T&D = Training and development

increased amounts in damages (Hor, 2012; Jenero & Galligano, 2003). A small number of organisations took compliance to new levels and identified an extension of their harassment policies to include protections for other groups, and identified issues of vilification and bullying throughout their policies and procedures.

Compliance was also an important consideration in addressing the issues of pregnancy and breastfeeding. Many organisations had policies specific to meeting the requirements of the legislation, but a small number had extended these to include further issues such as adoption and *in vitro* fertilisation requirements, while others ensured the policies in these areas were also available for either parent.

Correlation of the EEO approach and numbers of women and men

The data were examined using multiple regression analyses to ascertain any relationship between the policies and practices used and the numbers of men and women within the industry and in management. The only correlation for increasing numbers of women within the industry was a positive relationship with organisational size. A multiple regression controlling for size was performed with numbers of women in management as the DV and the approach undertaken by the organisations across the seven employment matters as the IVs. A second multiple regression analysis, controlling for size, was performed with numbers of men in management as the DV and the EEO undertaken by the organisations across the seven employment matters as the IVs. The model showed no relationship between the approach taken in implementing equality and diversity and the number of women in management (or the number of men in management). Further multiple regression analyses were run, also controlling for size, with the number of women and the number of men in supervision and operations as the DVs and the policies and practices undertaken by the organisations across the seven employment matters as the IVs. Results were similar, with one difference: the policies and practices undertaken to encourage equality in 'promotion and transfer' correlate to the increased numbers of men employed in supervision and operations.

A series of multiple regressions was undertaken, controlling for size, with the number of women and number of men in clerical and sales positions as the DV and the policies and practices undertaken by the organisations across the seven employment matters as the IVs. Results showed that the approach undertaken to encourage equality through policies in 'promotion and transfer' correlates to an increased number of women in clerical and sales positions in the construction industry. Results also showed that policies and practices undertaken to encourage equality in 'recruitment and selection,' 'promotion and transfer' and 'training and development' correlated with increased numbers of men in clerical and sales positions.

In our second study, we analysed the latest WGEA statistics for the construction industry. Women make up a high percentage of employees in traditional work areas in construction and are not significantly represented in management. Further, employers are not developing strategies for addressing equal pay, flexible work arrangements or primary carer leave to the same extent that employers in other industries are doing (see Table 9.3).

A total of 14 reports of the top 20 construction companies in Australia were available for analysis. Firms ranged in size from 68 employees to 13,000+ employees. (See Table 9.1b.)

Table 9.3 Construction industry profile for 2015 compared to total of all industries

Measure	Construction industry	Total of all industries
No. of reporting organisations	203	4,670
No. of employees	132,805	3,974,792
Percentage of women employed	16.2%	48.8%
Percentage of women CEO	2.7%	15.4%
Percentage of women as key management personnel	12.3%	27.4%
Percentage of women as senior managers	10.1%	33.%
Percentage of women in technical trades	2.2%	11.7%
Percentage of women in clerical and administration	77%	74%
Total remuneration pay gap for men and women	26.3%	24%
Employers that have an overall gender equality strategy	16.3%	20.6%
Employers that have set a target for gender composition of governing bodies	9.9%	16%
Employers with flexible work arrangement strategies	11.8%	14.6%
Employers that offer primary carer leave	23.2%	48.2%
Employers that offer full pay in addition to government scheme (maternity pay)	55.3%	80.9%
Employers that offer secondary carer leave	20.2%	38.4%

Table 9.1b Analysis of individual top construction organisation reports

Company	1	2	3	4	5	6	7	8	9	10	11	12	13	14
Total no. of employees	158	358	308	68	194	704	281	3038	5304	358	13,729	811	2296	734
% women employed	37%	40%	19.1%	16%	37%	19%	21%	13.7%	10.3%	11.7%	13.1%	7%	47%	7.6%
% women managers	25.7%	19.3%	14.2%	16%	22%	3%	6%	14%	11.3%	8.5%	15.6%	11.6%	38%	0%
Policies and practices that support gender equality	AD	AD	AD	Nil	Nil	EEO	AD	EEO	AD	EEO	EEO and MD	EEO	AD	Nil
Gender composition of governing bodies	EEO	Nil	EEO	Nil	Nil	Nil	Nil	Nil	Nil	Nil	Nil	EEO	Nil	Nil
Addressing the pay gap	Nil	Nil	Nil	Nil	Nil	Nil	Nil	Nil	Nil	Nil	AD	Nil	Nil	Nil
Flexible work and support for carers	AD	Nil	AD	Nil	Nil	Nil	AD	AD	AD	AD	EEO	AD	AD	Nil
Consultation with employees on issues concerning gender equality in the workplace	Yes/Committee	Nil	Representative group	Nil	Limited	Limited	EEO	EEO	Limited	Nil	EEO	Nil	EEO	Nil
Sex-based harassment/discrimination policies and practices	AD + annual training	Minimal	Minimal	Minimal	AD + training	AD + training	AD + training	AD + training	AD + training	AD + training	AD + training	AD + training	AD + training	Minimal

- **Gender equality indicator 1:** In the section for reporting on the occurrence of formal policies or formal strategies in place that SPECIFICALLY SUPPORT GENDER EQUALITY (capitals are specific to the reporting form) about recruitment, retention, performance management; promotion; talent identification; succession planning; training and development; resignations, etc., nine of the sample organisation, or 64 per cent, were barely compliant with the legislation. These organisations either reported nothing in any category or reported that they had policies in some or all of the areas with no strategies for implementation. Those who did report strategies, specifically organisations numbered 6, 8 and 12, reported strong EEO (gender specific programmes) and diversity programmes designed to address the disparity in numbers, mentor women, or appoint apprentices. For example, Company No. 6 reported: "We achieved our 50:50 male to female ratio for our graduate programme".

- **Gender equality indicator 2:** In the section for reporting on the gender composition of the governing board only three organisations (21 per cent) identified that they had a strategy for getting women on their boards. Company No. 1 reported "Measurable objective as per Diversity Policy is 'At least one female Board member' ". They reported one female on their board and were one of only two companies in this sample to have women on the board. Some organisations suggested it was not possible to have a policy or strategy because it was a family company or the shareholders elected the board, or they were a subsidiary of an international company, while others simply said "No" to such a policy or strategy.

- **Gender equality indicator 3:** In the section reporting on the equal remuneration between women and men only one organisation reported any activities, while some reported they had a policy, and most reported nothing at all! Many said that a remuneration review had occurred in previous years but reported that no subsequent activities had occurred. Organisation No. 12 indicated that they had created a pay equity strategy or action plan; identified the causes of the gaps; reviewed remuneration decision processes; analysed commencing salaries by gender to ensure no pay gaps; analysed performance rating to ensure there was no gender bias; examined performance pay to ensure there was no gender bias; and reported pay equity metrics to the board. They were a stand-out organisation in this category.

- **Gender equality indicator 4:** In the section reporting on the flexible working and support for employees with family and caring responsibilities five organisations (38.5 per cent) indicated that they were doing nothing in the work flexibility and caring considerations space and that these remained the premise of the individual through informal mechanisms. Eight organisations (57 per cent) reported on a range of work flexibility and caring considerations that were in the minimalist range with policies rather than strategies, and most activities in the informal range. One organisation reported on proactive strategies that could be considered EEO and diversity policies and practices with indicators that all these strategies were linked to formal processes for employees.

- **Gender equality indicator 5:** In the section reporting on consultation with employees on issues concerning gender equality in the workplace every organisation indicated that they did consult with their employees. However, we noted that most did not have any mechanisms for doing so. Only six or 43 per cent of the organisations indicated they consulted with a representative group of some kind. We identified these as proactive in their approach.

- **Gender equality indicator 6:** In the section reporting on sex-based harassment and Discrimination all organisations indicated that they had a standalone policy. Four, or 28.5 per cent of organisations, stated that they went no further (less than compliant), but others indicated the use of strategies including grievance processes and training of various types to support the policies for prevention of sexual harassment and discrimination.

Comparison of outcomes

Three attempts at different policy implementation requirements for developing equal opportunity in Australia have resulted in less than striking results and remarkably similar results. First, under the *Affirmative Action (Equal Opportunity for Women) Act 1986* organisations were encouraged through a prescriptive eight-step approach in HR programme development, with organisations reporting on programmes implemented. Organisational implementation was classified into four theoretical equity management structures reflecting different attitudes to distributive justice and organisational intervention. Some organisations took a traditional market-driven approach, ignoring the requirements of organisational intervention in addressing equal opportunity. Some organisations took an anti-discrimination approach, focusing on legislative compliance for equal and non-discriminatory treatment within programmes. Some organisations took an equal opportunity approach with gender-specific special measures programmes, and finally, there were organisations that undertook a gender diversity approach with gender-specific and gender-neutral programmes addressing both men and women particularly in the organisation of work. Equal treatment strategies specifically in the area of training and development were predictors of a significant increase in one indicator of the employment status of women, namely women moving into the top tier management level from the lower management level. Gender-specific special measures programmes were consistently the best predictor of increasing numbers of women managers overall and across all tiers of management.

Second, under the *Equal Opportunity for Women in the Workplace Act, 1999* organisations were required to strategically identify their approach to addressing seven employment matters. Implementation and resulting outcomes changed little. Industry differences were detected, with male-dominated organisations predominantly implementing equal opportunity through the equal treatment of men and women offering no significant change to the status quo in the representation of women in management or non-traditional work areas.

Third, under the *Workplace Gender Equality Act 2012* the requirements have changed to an inclusive outcome focus with organisations required to report the statistical changes (along with implementation along six gender equity indicators) for both men and women with a view to benchmarking within and between industries. Though this new Act has limited years of reporting outcomes, equal opportunity measures of implementation in the male-dominated industry of construction have changed little with limited strategic or operational implementation of special measures to address inequity, and women remain the outsiders when it comes to jobs in management or non-traditional areas of work.

Implications for practice and research

The effectiveness of policy development and implementation has become a driving force for managing equality and inclusion globally with evidence that legislative intervention brings significant results (Thornton, 1990). This study focuses on the ways three legislative changes have impacted on implementation approaches to equity management in the construction industry in Australia. Three attempts at different policy implementation have resulted in less than striking outcomes and remarkably similar results. The chapter asks the question whether there have been any changes for women in these organisations, and the answer is not many. Overall, the results continue to suggest that inequality regimes (Acker, 2006) continue to thrive in organisations, particularly in male-dominated industries.

National legislation which focuses on organisations addressing gender equity is an essential basis for action as it sets national standards and expectations. This study of organisations in the construction industry shows that legislation which requires reporting on gender equity measures has produced little change in outcomes. Some industry sectors, such as the university sector, have been very active in the development of policies, resulting in a significant proportion of universities achieving national recognition, with the consequence that the expectation across this sector is that all universities will have developed leading policies and practices (WGEA, 2016d). In contrast, the construction industry has engaged with the legislation at a minimal level focused on legal compliance. We have noted (French & Strachan, 2007) that this approach does more to maximise labour market participation, moving men and women in and out of organisations, than it does to assist in substantive equity in pay, management opportunities, career development or the management of other areas of life (e.g. caring) as questioned by Watt (2009). This level of engagement has remained persistent since the 1980s, and the question now is how to move these organisations from minimal engagement.

Consideration needs to be given to the means and details of reporting as this will inevitably influence organisational thinking. The implementation of minimum industry standards for organisations with more than 500 employees under the 2012 Act is essential, especially for industries such as construction where the level of engagement is low. Special attention needs to be focused on male-dominated industries, especially construction, and assistance and guidance be given – until certain practices become the norm. Consideration needs to be given to the implementation of different measures, and positive goals and actions are required.

While this study is longitudinal in nature, the lack of consistency across the reporting instruments is not helpful. However, this does not limit the value of the research as ongoing work in equality and inclusion management unique to the construction industry. The typology used is built on a solid statistical base of more than 1,900 organisations and assisted our investigation into the various implementations of treatment in the name of equity. The importance of this extended research is that it enables us to pay particular attention to the specifics of work organisation where gender is enacted (Robinson et al., 2005). All the reporting instruments through each legislative change have continued to contain employment data by sex and job type linked to each organisation and accurate information on organisational policies related to issues of equal opportunity in the workplace, allowing us to answer the call "…to develop better comparative organizational data that enable us to view class-gender-ethnic dynamics and outcomes across and within workplaces" (Robinson et al., 2005: 7).

Summary

Internationally the employment of women within the construction industry remains woeful, despite decades of research, anti-discrimination and equal opportunity legislation across many jurisdictions and the pressure of best practice in diversity management in practice. Women remain both vertically and horizontally challenged in their positions within construction organisations with organisational confirmation that pay equity is almost non-existent and currently not receiving attention. Australia provides an interesting and valid national case study regarding equity management through the implementation of equity legislation which forces organisations to engage in the design and implementation of policies and practices designed to overcome inequity in a range of areas. Despite the development of a national approach to equity management, industry differences in implementation are evident, with male-dominated industries engaging with equity management through minimalist strategies of equal treatment that are designed to treat unequal people equally and are not substantively changing the status quo.

When considering whether this system of developing legislative approaches can be improved within intractable industries, we note that other industries are engaging with the legislation to a higher degree with substantive change as an outcome. For this to occur in construction, the system may need to provide specific industry-based evidence and mandatory initiatives within an industry framework with national expectations for change.

References

Acker, J. (2006), Inequality regimes: Gender, class and race in organizations, *Gender and Society*, Vol. 20(4), pp. 441–464.

Australian Government. (1999), *Equal Opportunity for Women in the Workplace Act 1999*.

Australian Mines and Metals Association (AMMA). (2016) [Online]. Available at: www.miningoilgasjobs.com.au/construction/all-you-need-to-know/top-construction- companies-in-australia.aspx Date of access: 2 September 2016.

AMMA *see* Australian Mines and Metals Association.

Bacchi, C. (2000), The seesaw effect: Down goes affirmative action, up comes managing diversity, *Journal of Interdisciplinary Gender Studies*, Vol. 5(2), pp. 64–83.

Cartwright, S. & Gale, A. (1995), Project management: Different gender, different culture? A discussion on gender and organizational culture part 2, *Leadership & Organizational Development Journal*, Vol. 16(4), pp. 12–16.

Catalyst. (2015), *Women in male-dominated industries and occupations* [Online]. Available at: www.catalyst.org/knowledge/women-male-dominated-industries-and-occupations Date of access: 2 November 2016.

Crawford, L., French, E. & Lloyd-Walker, B. (2013), From outpost to outback: Project career paths in Australia, *International Journal of Project Management*, Vol. 31(8), 1175–1187.

Crompton, R. & Harris, F. (1999), Employment, careers, and families: The significance of choice and constraint in women's lives. In Crompton, R. (ed.), *Restructuring gender relations and employment: The decline of the male breadwinner*, Oxford University Press, Oxford, pp.128–149.

Dainty, A., Bagilhole B. & Neale, R. (2001), Male and female perspectives on equality measures for the UK construction sector, *Gender in Management*, Vol. 16 (5/6), pp. 297–304.

Department of the Prime Minister and Cabinet. (1984), *Affirmative action for women: Volume 1: A policy discussion paper*, Australian Government Publishing Service (AGPS), Canberra.

Ellison, L. (2001), Senior management in chartered surveying: Where are the women? *Gender in Management*, Vol. 16(5/6), pp. 264–278.

English, J. & Hay, P. (2015), Black South African women in construction: Cues for success, *Journal of Engineering, Design and Technology*, Vol. 13(1), pp. 144–161.

EOWA (2006), Employer of Choice for Women [Online]. Available at: www.eeo.gov.au/EOWA_Employer_of_Choice_for_Women.asp Date of access: January 2006.

Fielden, S., Davidson, M., Gale, A. & Davey, C. (2001), Women, equality and construction, *The Journal of Management Development*, Vol. 20(4), pp. 293–304.

Fowler, F.J. Jr. (1988), *Survey research methods* (rev. edn), Sage, Thousand Oaks, CA.

French, E. (2001), Approaches to equity management and their relationship to women in management, *British Journal of Management*, Vol. 12(4), pp. 267–285.

French, E. & Strachan, G. (2007), Equal opportunity outcomes for women in the finance industry in Australia: Evaluating the merit of EEO plans, *Asia Pacific Journal of Human Resources*, Vol. 43(3), pp. 314–332.

French, E. & Strachan, G. (2009), Evaluating equal employment opportunity and its impact on the increased participation of men and women in the transport industry in Australia, *Transport Research Part A: Policy and Practice*, Vol. 43A(1), pp. 78–89.

French, E. & Strachan, G. (2015), Women at work! Evaluating equal employment policies and outcomes in construction, *Equality, Diversity and Inclusion: An International Journal*, Vol. 34(3), pp. 227–243.

Gale, A. (1994), Women in non-traditional occupations, *The Construction Industry Gender in Management*, Vol. 9(2), pp. 3–14.

Gurjao, S. nd, *Inclusivity: The changing role of women in the construction workforce*, The Chartered Institute of Building, Ascot, Berkshire, UK [Online]. Available at: www. ciob.org/sites/default/files/CIOB%20research%20-%20The%20Changing%20 Role%20of%20Women%20in%20the%20Construction%20Workforce.pdf Date of access: 2 November 2016.

Harris, H. (2001), Content analysis of secondary data: A study of courage in managerial decision making, *Journal of Business Ethics*, Vol. 34(3/4), pp. 191–198.

Hor, J. (2012), *Managing workplace behaviour*, Commerce Clearing House (CCH), Sydney.

Jenero, K.A. & Galligano, M.L. (2003), Courts continue to emphasize importance of policy development and training, *Employee Relations Law Journal*, Vol. 28(4), pp. 113–124.

Kirton, G. & Greene, A. (2005), *The dynamics of managing diversity* (2nd edn), Butterworth-Heinemann, Elsevier.

Konrad, A.M. & Linnehan, F. (1995), Formalized HRM structures: Coordinating equal employment opportunity or concealing organizational practices? *Academy of Management Journal*, Vol. 18(3), pp.787–820.

Krippendorff, K. (1980), *Content analysis*, Sage, Beverly Hills, CA.

Powell, A. & Sang, K. (2015), Everyday experiences of sexism in male-dominated professions: A Bourdieusian perspective, *Sociology*, Vol. 49(5), pp. 919–936.

Randstad (2015) Women in the UK construction industry in 2016 [Online]. Available at: www.randstad.co.uk/women-in-work/women-in-the-uk-construction-industry-in-2016. pdf Date of access: 17 March 2017.

Robinson, C.L., Taylor, T., Tomaskovic-Devey, D., Zimmer, C. & Irvin, M.W. (2005), Studying race or ethnic or sex segregation at the establishment level, *Work and Occupations*, Vol. 32(1), pp. 5–38.

Strachan, G. & Burgess, J. (1998), The 'family friendly' workplace: origins, meaning and application at Australian workplaces, *International Journal of Manpower*, Vol. 19(4), pp. 250–265.

Strachan, G., Burgess, J. & Sullivan, A. (2004), Affirmative action or managing diversity – What is the future of equal opportunity policies in organisations, *Women in Management Review*, Vol. 19(4), pp. 196–204.

South Australian College of Advanced Education and Affirmative Action Agency (Australia) (1990) *Report to the Affirmative Action Agency*. Adelaide, South Australia: The College.

Sutherland, C. (2013), Reframing the regulation of equal employment opportunity: The Workplace Gender Equality Act 2012 (Cth), *Australian Journal of Labour Law*, Vol. 26(1), pp. 102–111.

Thornton, M. (1990), *The liberal promise – Anti-discrimination in Australia*, Oxford University Press, Sydney.

Union of Construction, Allied Trades and Technicians (UCATT), *The trade union for construction workers* [Online]. Available at: www.ucatt.org.uk/women-construction [Accessed 9 November 2016].

UCATT *see* Union of Construction, Allied Trades and Technicians

United States Department of Labor [Online]. Available at: www.osha.gov/doc/topics/ women/

Watts, J. H. (2009), 'Allowed into a man's world' meanings of work-life balance: Perspectives of women civil engineers as 'minority' workers in construction, *Gender, Work and Organization*, Vol. 16(1), pp. 37–57.

Warren, C. & Antoniades, H. (2015), Deconstructing the glass ceiling: Gender equality in the Australian property profession, *Property Management*, Vol. 34(1), pp.29–43.

Workplace Gender Equality Agency (WGEA). (2016a), *Gender composition of the workforce: By industry* [Online]. Available at: www.wgea.gov.au/sites/default/files/Gender%20 composition-of-the-workforce-by-industry.pdf Date of access: 2 November 2016.

Workplace Gender Equality Agency (WGEA). (2016b), *Gender segregation in Australia's Workforce* [Online]. Available at: www.wgea.gov.au/sites/default/files/20160801_Industry_ occupational_segregation_factsheet.pdf Date of access: 2 November 2016.

Workplace Gender Equality Agency (WGEA). (2016c), *What are minimum standards?* [Online]. Available at: www.wgea.gov.au/minimum-standards/what-are-minimum-standards Date of access: November 2016.

Workplace Gender Equality Agency (WGEA). (2016d), *Employer of choice for gender equality* [Online]. Available at: www.wgea.gov.au/lead/employer-choice-gender-equality-0 Date of access: 2 November 2016.

Workplace Gender Equality Agency (WGEA). (2016e), *Public reports* [Online]. Available at: www.wgea.gov.au/report/public-reports Date of access: 2 November 2016.

WGEA *see* Workplace Gender Equality Agency.

Worrall, L., Harris, K., Stewart, R., Thomas, A. & McDermott, P. (2010), Barriers to women in the UK construction industry, *Engineering Construction and Architectural Management*, Vol. 17(3), pp. 268–281.

10 Work-family conflict within civil engineering

The role of family supportive policies (FSPs) and work-family culture

Valerie Francis

Live your life as though your every act were to become a universal law

(Immanuel Kant)

Background

The 2013 American-British movie, *Locke*, depicts some of the work-family issues faced by civil engineers, particularly those in construction. Tom Hardy, the only actor who appears in the movie, is depicted driving alone in his car on his way to London for the birth of his illegitimate child. The movie chronicles his 36 phone calls with his boss, wife, son, former mistress and second-in-charge as he desperately juggles the complex planning of an imminent concrete pour, breaking the news of his infidelity (whilst away on a previous project) to his family and his attempts to reassure the woman giving birth to his child. In the end, he successfully puts into place the resources for the pour but is fired from his job and kicked out of his home. While this is a very dramatic depiction, it does highlight the themes of long work hours, excessive work responsibilities and relocation for work that echo many of the precursors to work-family conflict in work-family literature, themes that are well understood in the civil engineering profession (Eby et al., 2005; Lingard & Francis, 2009).

But why has this topic gained such prominence? Most would ascribe this to demographic changes within society, both regarding labour force participation and an ageing population, as well as changing societal expectations. The employment rates of women have risen, with dual-earner couples being widespread throughout the workforce, and single parents are more highly represented within work (Australian Bureau of Statistics, 2009). Along with increasing life expectancy rates, balancing the care of children and elderly relatives in conjunction with work responsibilities has become more prevalent, especially for women (Page et al., 2009). Added to this has been a shift in both male and female work and life expectations. For women, there has been an increase in the social and financial importance of paid work, whereas for men it has been an increase in a desire for greater involvement in their children's daily lives and a concomitant questioning of the dominance of work within their lives (Lewis & Sussman, 2013).

In Australia in 2002, the then Prime Minister, John Howard, even coined a new term, "barbecue stopper", to refer to the effect the discussion of the challenges of achieving work-life balance had on one of the most iconic symbols of the Australian lifestyle (Browne, 2010). However, despite annual, then biennial reviews, of work-life issues in Australia since 2007, the most recent work-life index found that no real change in index scores had occurred, despite significant work-force changes. The authors stated that "Work-life interference remains a persistent challenge in Australia despite some changes in childcare, parental leave, and employment law in the past two decades" (Skinner & Pocock, 2014: 1). These trends are not confined to Australia, where this research was conducted, but are repeated within most developed countries throughout the world.

In this chapter, the work-life experiences of Australian civil engineers were examined. In particular, the level of family supportive policies (FSPs) available, predictors of an organisational culture supportive of employee work-family balance and the effect of work-family culture on work-family conflict were explored. As women are in the minority in this profession, and women still bear a greater burden of family responsibilities, some gender differences were also considered. While civil engineering is not alone in experiencing work-life balance issues, it is of interest as male-dominated professions often lag behind in the adoption of FSPs and men in male-dominated professions are less likely to request altered work arrangements for family reasons, so may suffer more than the general population (Skinner & Pocock, 2014).

Conceptual foundation

Work-family conflict

Over 30 years ago Greenhaus and Beutell (1985: 77) described the work-family conflict as a "...form of inter-role conflict in which the role pressures from the work and family domains are mutually incompatible in some respect". They explain how participation in the family or home role is more challenging owing to the involvement in the work role and vice versa. Research since then has clearly identified that there are many negative consequences, particularly from the work domain to the family or home domain. This negatively impacts individuals, their families, and their organisations and similar results have been demonstrated in construction (Allen, 2012; Lingard & Francis, 2009).

A 2005 meta-analysis of work-family conflict antecedents found that work-related factors caused the greatest work interference with family, with demographic variables having little effect (Byron, 2005).(Byron (2005) also demonstrated that family and non-work variables were related to family interference with work (or family-work conflict).) In particular, workload such as work hours was related to higher levels of work-family conflict whereas work, job flexibility and supportive co-workers and supervisors were linked to lower levels of work-family conflict. More recent studies have found that greater control at work was found to be related to lower levels of work-family conflict (Lapierre & Allen. 2012). Both role conflict and role ambiguity, which arise from opposing and conflicting institutional demands, have been found to cause workplace stress, depression and anxiety (Kahn et al., 1964: Lambert & Lambert, 2001; Schmidt et al., 2014; Soltani et al., 2013).

It is now also well understood that higher levels of work-family conflict are associated with many adverse organisational and personal outcomes, such as increased turnover intentions, burnout and stress, as well as lower levels of job and life satisfaction (Jones et al., 2013; Edwards & Rothbard, 2000; Eby et al., 2005; Lingard & Francis, 2009). In project-based industries such as construction and civil engineering, making use of the fluctuations in workload by building in recovery opportunities can improve work-life balance (Lingard et al., 2010). Finding further ways to reduce work-family conflict of employees should be an important business strategy, with many organisations responding by offering family-friendly benefits or family supportive policies (FSPs) such as flexible work arrangements, part-time employment options, financial support for the care of dependents and well-being programmes.

Work-family culture

Many businesses understand the significance of developing and nurturing a work-place culture that is supportive of employees' desire to have better work-life balance. Work-family culture refers to the "…shared assumptions, beliefs, and values regarding the extent to which an organization supports and values the integration of employees' work and family lives" (Thompson et al., 1999: 394). This culture comprises both supportive work practices as well as the availability of FSPs. A positive work-family culture has been found to be related to increased job satisfaction, as well as lower work-family conflict and turnover intentions; however, a negative culture undermines the use of FSPs (Thompson et al., 1999; Ngo et al., 2009; Lewis, 2001; Andreassi & Thompson, 2008).

While previous research (Michel et al., 2011) has demonstrated that stressors such as role conflict and role ambiguity contribute to work-family conflict, research by Allen (2001) demonstrated that this might not be a simple relation-ship, with the work environment playing a significant part in the usage of family-friendly policies or FSPs. FSPs include policies relating to flexibility in when and where work can be conducted, schemes which support the integration of work and family life and financial support for work and life integration (e.g. support for childcare or fitness). While these are normally developed to suit a range of employees, some rarely access them owing to beliefs they may be viewed nega-tively by management. Studies have had mixed results about the benefit of work flexibility on work-family conflict. Thompson et al. (1999) suggest this may be due to the level of support available in the organisation.

A recent study found a significant relationship between the use of FSPs and work-family culture (De Sivatte et al., 2015). Also, supervisors supportive of employees' family responsibilities positively affect staff job satisfaction, organisa-tional commitment, as well as lower turnover intentions (Rofcanin et al., 2016). The provision of FSPs, along with a reduction in work stressors and a supportive workplace, may influence the work-family culture of the organisation, which in turn affects the level of work-family conflict.

This research sought to explore how the availability of FSPs and work-family culture was related to the work-family conflict. From the literature reviewed some hypothesised relationships were developed. In particular, it was considered that a supportive work-family culture is shaped not only by the FSPs available but also issues such as work hours, support from supervisors and colleagues and the levels of control the employees have over work (specifically work flexibility, role ambiguity, and role conflict). It was also considered that the work-family culture of an organisation then, in turn, affects the level of work-family conflict experienced by employees.

The notion that FSPs, supportive work practices and better control over work contribute to work-family culture and work-family conflict is based on both organisational support theory and resource theory (Eisenberger et al., 1986; Goode, 1960; Edwards & Rothbard, 2000). As resources such as time and energy are finite, when work demands are greater, fewer resources are left for the non-work domain – thus work-family conflict develops. FSPs such as flexibility, along with better or more supportive work practices, provide either additional resources or greater discretion on when, how and where work is completed (Allen et al., 2013).

Women (and men) in civil engineering

Why analyse the data by gender? A report published in 2016 by Engineering UK summed up the situation regarding women in engineering in the following manner: "The engineering community must recognise and address the fact that, despite numerous campaigning initiatives over the past 30 years, there has been no significant advance in the diversity or make-up of the sector. In particular, the gender participation of women into engineering must change" (Engineering UK, 2016: IX). Women are still in the minority within civil engineering, and work-family issues are often noted as a reason for their leaving the profession (Wilkinson, 2006; Watts, 2009). In the UK, for instance, only 15 per cent of engineering degrees were awarded to women in 2013/14, and 17 per cent of all civil engineering students are women (Engineering UK, 2016; Peers, 2016). The percentage of women working in engineering in the UK in 2014 was around 9 per cent, and in Australia, 11.8 per cent of the engineering workforce were women in 2011 (Peers, 2016; Kaspura, 2015).

In addition, it is important to recognise that engineering is not a profession that has a high uptake of part-time work (13.3 per cent vs 32.5 per cent in all occupations) and it is not just female engineers who seek alternative work arrangements. While the proportion of women working part-time is around 27 per cent, "…numerically there were almost three times as many men employed part-time than women" (Kaspura, 2015: 11). Interestingly "…the largest group of women not in the labour force, 42.2 per cent, were aged from 25 to 39 years, often regarded as child rearing years" and "fewer women are retained in engineering than men" (Kaspura, 2015: 7, 10). This data indicates that retention of women is problematic in engineering and the work-family relationship is an area worthy of attention for both men and women.

Research aims and method

The aim of this research was to investigate the levels of work-family conflict experienced by civil engineers, the availability of FSPs and the work-family culture within civil engineering organisations. In particular, this research examined the predictors of work-family culture, as well as the effect of a work-family culture on work-family conflict. Some gender differences were also investigated. This research was conducted in Australia.

Methodology, procedure and participants

A cross-sectional correlational field study was adopted as the purpose of the study was to examine the extent to which variation in one or more characteristics is related to differences in other characteristics (Leedy & Ormrod, 2016). The sample was recruited with the help of an Australian professional organisation. Invitations to participate were sent to 500 male and 500 female civil engineers, who were randomly selected by the organisation. The engineers were aged between 25 to 55 years as it was considered they best suited the nature of the study. The final sample was 194 and consisted of 87 males and 107 females; 74.7 per cent were partnered, and 53.1 per cent had children. The sample was adequate for the statistical tests being conducted. The average age of the sample was 36.7 years, and the average working hours per week was 45.2 (SD = 9.9). A larger proportion of the sample were employed in the public sector than would be expected and the sample performed a range of roles from mainly technical to main management. Further information on the sample can be found in Table 10.1.

Table 10.1 Demographic characteristics of the sample

	N	%		N	%
Gender			**Average no of employees**		
Male	87	44.8	1 – 19	29	15.1
Female	107	55.2	20 – 49	24	12.5
Family structure			50 – 199	21	10.9
Couple with children	97	50.0	200 – 999	54	28.1
Couple no children	48	24.7	1000 or more	64	33.3
Single parent	9	4.6	**Employment sector**		
Single person	36	18.6	Public	70	36.1
Other	4	2.1	Private	124	63.9
No of children			**Work type**		
0	91	46.9	Mainly technical	38	19.7
1	23	11.9	75% technical/25% managerial	51	26.4
2	54	27.8	50% technical/50% managerial	36	18.7
3	23	11.9	75% management/25% technical	34	17.6
4	3	1.5	Mainly management	34	17.6

Measures

The questionnaire was divided into several sections and along with demographic information, some of which has already been noted, perceptions were gathered using several existing measures. These are summarised below.

The work-family culture was measured via a 20-item scale (Thompson et al., 1999) that investigated three main areas, namely managerial support, career consequences, and organisational time demands. Respondents were asked to decide to what extent the statements characterised their current organisation using a scale ranging from 'strongly disagree' (1) to 'strongly agree' (7). It included items such as: "In general, managers in this organisation are quite accommodating of family-related needs".; "To turn down a promotion or transfer for family-related reasons would adversely hurt one's career progress in this organisation. (R)"; and "To get ahead in this organisation, employees are expected to work more than 50 hours a week, whether at the workplace or home. (R)". The coefficient alpha reliability for each sub-construct was .85, .83 and .81 respectively. A higher score was more indicative of a more supportive culture.

Work-family conflict was measured using a five-item, seven-point scale ranging from one (1) (for strongly disagree) to seven (7) (for strongly agree), developed by Netemeyer, Boles, and McMurrian (1996). The coefficient alpha reliability for the scale was .82. Items were summed and averaged with higher scores indicating a greater degree of conflict. A sample item was "The demands of my work interfere with my home and family life".

Supervisor support was measured via a three-item satisfaction scale developed by Hackman and Oldham (1975) as part of the Job Diagnostic Survey. Respondents were asked how satisfied or dissatisfied they were (one [1] for 'extremely dissatisfied' to seven [7] for 'extremely satisfied'). A sample item was "The amount of support and guidance I receive from my supervisor". The alpha reliability was .78. *Colleague support* was measured in a similar manner but using a single item to determine their satisfaction with the level of support participants received from their partners.

Work flexibility was measured using a three-item five-point Likert scale adapted from Hill, Hawkins, Ferris, and Weitzman (2001) which focused on how much flexibility they had in deciding where and when their work was done, as well as how and what work was to be undertaken. As complete flexibility was rated one (1), the item scoring was reversed so that a higher number indicated greater flexibility. The alpha reliability was .64 which was lower than .7 recommended by Pallant (2011).

Role Ambiguity and *Role Conflict* were measured using six-item and eight-item scales developed by Rizzo, House, and Lirtzman (1970). The scales used a seven-point scale ranging from 'very false' to 'very true', and the alpha reliability was .81 for each scale. Sample items for role ambiguity include "I know exactly what is expected of me (R)" and "I feel certain about how much authority I have (R)". Role conflict sample items include "I receive incompatible requests from two or more people" and "I do things that are apt to be accepted by one person and not accepted by others". A higher score was indicative of greater ambiguity or conflict in their role.

Family supportive policies (FSPs) were derived from a range of sources, including company websites and scholarly articles (e.g. Budd & Mumford, 2006; Gray & Tudball, 2003; Bloom, Kretschmer & Van Reenen, 2011). In total 15 FSPs were selected and included a range of flexible work practices, supportive practices, and financial family support. Participants were asked to indicate availability (available = 1 and not available = 0). If they did not know whether a policy existed it was also coded as zero. The FSP score was a sum of available policies (15 in total), where a higher number was indicative of more policies. In addition to gauging the availability of FSPs, respondents were also asked about the usefulness of the FSP on a three-point scale ('I use it now', 'I may use it in the future' and 'I would not use it').

Analysis

Data was analysed using IBM SPSS (version 22), and the results are presented in Tables 10.1 to 10.4. Bi-variate correlation analyses are presented in Table 10.2 and were conducted to assess the degree to which one variable is linearly related to another, thereby determining the direction and strength of linkages between variables. Multiple regression analyses are presented in Tables 10.3 and 10.4 and were used to determine the main effects of the hypothesised relationships. Cohen's f^2 is the best-known way to measure the effect size in multiple regressions with f^2 effect sizes of 0.02, 0.15 and 0.35 respectively termed small, medium and large (Cohen, 1988). An alternative way to consider effect sizes is to

Table 10.2 Means, SD and bivariate correlations of work-family variables

	Mean	SD	Work-family conflict	Work-family culture		
				Managerial support	Career conseq's	Org. time demands
Work-family conflict	4.44	1.39	1			
Managerial support	4.65	0.95	−.389**	1		
Career consequences	4.35	1.18	−.321**	.663**	1	
Org. time demands	4.17	1.39	−.513**	.628**	.592**	1
FSPs available	5.94	3.23	−.058	.340**	.215**	.079
Work hours/week	45.24	9.93	.371**	−.187**	−.127	−.308**
Work flexibility	2.81	.95	−.109	.278**	.216**	.170*
Supervisor satisfaction	5.19	1.27	−.100	.229**	.228**	.165*
Colleague satisfaction	5.30	1.31	−.245**	.304**	.327**	.258**
Role ambiguity	2.98	.99	.136	−.245**	−.258**	−.197**
Role conflict	3.96	1.17	.340**	−.292**	−.293**	−.395**

$*p < .05, **p < .01, ***p < .001$

investigate β values. For multiple regressions, Keith (2006) considers the effect size as meaningless when β is below .05, small but meaningful when β is above .05, moderate when β is above .1, and large when β is above .25. Both effect sizes were considered.

Results and discussion

Overall, the results indicate that civil engineers work relatively long hours per week, averaging 45.2 but ranging from 7 to 75, along with moderate levels of work-family conflict and only a slightly supportive work-family culture ('neutral' was 4 with 7 'strongly agree'). The bi-variate correlation reveals that engineers with higher levels of work-family conflict also reported significantly poorer work-family culture and higher work hours. While the availability of FSPs and work flexibility was not related to work-family conflict, they were significantly related to different elements of work-family culture. For FSPs, this was managerial support and career consequences, whereas for work flexibility this was managerial support and career consequences. Both supervisor and collegial support were positively correlated with all aspects of work-family culture. However, only collegial support was negatively related to the work-family conflict. Similarly, while role ambiguity and role conflict were negatively related to work-family culture, only role conflict was positively related to the work-family conflict. These indicate that collegial support has a large role in the work-family experience of civil engineers, and role ambiguity and role conflict feature in workplaces which have a poorer work-family culture.

Regression analyses

Four multiple regressions were conducted: the first three to assess the ability of FSP availability, work flexibility, work hours, collegial and supervisory support, and role ambiguity, and role conflict to predict the various aspects of the work-family culture construct (managerial support, career consequences and organisational time demands). These are presented in Table 10.3. The final regression (Table 10.4) assessed the ability of work-family culture to predict work-family conflict. In all cases, preliminary analyses were conducted to ensure that the assumptions of normality, linearity, multi-collinearity and homoscedasticity were not violated.

The first multiple regression was used to assess the ability of seven variables to predict the managerial support aspect of work-family culture. Using the enter method, a significant model emerged, $F(7, 146) = 7.675$, $p < 0.001$. The total variance in managerial support explained by the variables overall was 26.9 per cent. Cohen's f^2 was used to assess the effect size, with f^2 of .37 indicative of a large effect (Cohen, 1988). The variable recording the highest beta value and therefore the strongest influence on the managerial support was FSP availability ($\beta = .256$). In this regression, the effect size of the beta for FSP availability would also be considered as large (Keith, 2006). Collegial support would be regarded as

Table 10.3 Multiple regression of work-family culture

Variables	Managerial support			Career consequences			Org. time demands		
	B	SE B	β	B	SE B	β	B	SE B	β
(Constant)	4.907	.733		4.612	.941		6.599	1.079	
FSP availability	.075	.024	.256**	.047	.030	.130	-.002	.035	-.005
Work flexibility	.079	.083	.079	.100	.106	.080	.085	.122	.058
Work hours	-.010	.007	-.107	-.006	.009	-.051	-.031	.011	-.220**
Supervisory support	-.056	.076	-.075	-.097	.097	-.105	-.139	.111	-.127
Collegial support	.160	.065	.221*	.248	.083	.275**	.256	.096	.241**
Role ambiguity	-.129	.084	-.134	-.164	.108	-.137	-.156	.124	-.111
Role conflict	-.159	.065	-.196*	-.219	.083	-.217**	-.360	.095	-.303***
R^2		.269			.217			.260	
F		7.675***			5.775***			7.330***	

*$p < .05$, **$p < .01$, ***$p < .001$

Table 10.4 Multiple regression of work-family conflict

Variables	B	SE B	β
(Constant)	7.005	.435	
Managerial support	-.181	.132	-.124
Career consequences	.034	.103	.029
Organisational time demands	-.451	.084	-.452***
R^2		.271	
F		23.274***	

*$p < .05$, **$p < .01$, ***$p < .001$

approaching large while role conflict would be considered as small, approaching moderate. These findings suggest that FSP availability and collegial support have a large/considerable effect on managerial support.

The second multiple regression was used to assess the ability of seven variables to predict the career consequences' aspect of work-family culture. Using the enter method, a significant model also emerged, $F(7, 146) = 5.775$, $p < 0.001$. The total variance in career consequences explained by the variables overall was 21.7 per cent. The Cohen's f^2 was .28, indicative of a medium effect (Cohen, 1988). The variable recording the highest beta value was collegial support ($\beta = .275$) and the effect size of the beta would be considered as significant (Keith, 2006). Role conflict was also significant, and its effect size would be regarded as moderate but approaching large. These findings suggest that collegial support and role conflict have a large/ significant? effect on career consequence.

The third multiple regression was used to assess the ability of seven variables to predict the organisational time demands aspect of work-family culture. Using the enter method, another significant model emerged, $F(7, 146) = 7.330$, $p < 0.001$. The total variance in organisational time demands explained by the variables overall was 26.0 per cent. The Cohen's f^2 was .35, indicative of a large effect (Cohen, 1988). The variable recording the highest beta value and therefore the strongest influence organisational time demands was role conflict ($\beta = .303$), and the effect size of the beta would be considered as large (Keith, 2006). Work hours and collegial support were also significant, and their effect sizes would be considered moderate but approaching large. These findings suggest that role conflict, collegial support, and work hours have a large/considerable? effect on organisational time demands.

The final regression, presented in Table 10.4, was used to assess the ability of work-family culture to predict work-family conflict. Again using the enter method, a significant model emerged, $F(3, 188) = 23.274$, $p < 0.001$. The total variance in work-family conflict explained by work-family culture was 27.1 per cent. The Cohen's f^2 was .37 indicative of a large effect (Cohen, 1988). The only significant variable had a beta value of .452, and the effect size would be considered as large (Keith, 2006). These findings suggest that organisational time demands have a large effect on work-family conflict. Managerial support and career consequences had no significant effect. However, it should be noted that the three sub-constructs of work-family culture are all significantly correlated.

Gender differences

Gender differences were apparent in the perceived usefulness of FSPs, with females much more supportive of FSPs compared to their male colleagues. This is especially true for policies about children and aged dependants, such as opportunities to work part-time to care for children, additional leave during school holidays and provision of child care assistance. This may have been related to the fact that the male sample was on average five years older than the female sample (so some perhaps past childcare needs) or within more traditional family structures with less need for child care provisions. However, this reticence extended beyond child-related policies to use of assistance programmes for family problems, eldercare information, and wellness programmes. Interestingly, men were as likely as women to want to access work phones for family reasons as well as seeking more flexible work hours. However, this flexibility did not extend to where they were physically located for their work. Women expressed a much greater desire to be able to work from home or telecommute. Finally, men expressed little desire to work part-time, either now or in the future.

In addition, the hours involved in work and home activities for the male and female participants were compared, taking into account parental status. The average weekly work hours, hours spent on care of dependants and hours dedicated to household chores and maintenance are shown in Figure 10.1, along with partners' (spouse) weekly work hours. For the fathers in the sample, the total hours spent on the three named activities was 65.9 hours, with their partners working 32.8 hours on average. For the male respondents who were not fathers, their work hours and household hours were similar and, unsurprisingly, their hours spent caring for dependants were lower, resulting in around 60.6 hours spent on the same three activities. Their partners worked on average 43.2 hours, which was ten hours more than for the sample of fathers' partners.

Females who are not mothers have a very similar profile to the male engineers without children, averaging 60.1 hours on the three activities, with their

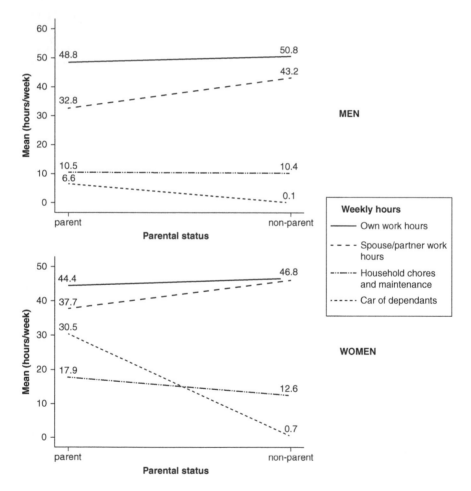

Figure 10.1 Weekly work and non-work hours of men and women by parental status

partners working slightly longer hours. However, the profile for civil engineers who were mothers was quite different. The total hours for the same three activities were 86.1, and their partners worked 44.4 hours per week – similar to their female and male colleagues without children. The work hours of partners of fathers who were civil engineers averaged five hours less per week than those who were mothers and civil engineers.

Fathers and mothers had similar levels of work-family conflict and women appear to mitigate their work-family conflict by working less. Interestingly, household weekly work hours for the male engineers were 81.6 compared with 82.1 for the female engineers. This may be why the work-family conflict scores for the parents were not statistically significantly different. Interestingly, while the number of FSPs available was similar for males and females, it was the women in the group who were more likely to use them than their male colleagues. Men in civil engineering appear more conservative in their adoption of FSPs. While they have similar levels of work-family conflict and a similar number of FSPs available to them as their female counterparts, they were less likely to access them. This may be because they have more traditional family structures based on the male breadwinner model and the fact the mothers in the sample have significant caring responsibilities.

Summary

Considerable social and demographic changes, not only in the composition of the workforce but also in the roles and expectations of men and women, have led to a greater number of workers having considerable non-work responsibilities. Many features of the civil engineering profession and the construction sector impact negatively on family lives, including long work hours, work inflexibility, high job demands, job insecurity and frequent relocation for work. As it is the dominant group which shapes the workplace culture, the civil engineering workplace has a prevailing masculine culture (Faulkner, 2009). This culture, which is based on shared formal and informal understandings, is based on the male, bread-winner model, assuming long work hours and 'loyalty above other' aspects of life. Productivity alone is not enough with a continual presence at work being valued.

However, this model no longer suits many men, particularly those who are younger and want to engage more actively in family life, as well as those who care for both children and elderly parents. In addition, as women increase their participation in the profession, companies employing civil engineers will need to review their policies, practices, and cultures. However, this cultural change transpires slowly in many organisations. As Watts (2009) observed, women in civil engineering are typically pressured into accepting masculine norms such as presenteeism and making themselves constantly available, rather than the culture changing for them.

In this chapter, the predictors of an organisational culture supportive of work-family balance were explored, as well as the influence of that culture on work-family conflict. Previous research demonstrates the positive attributes of a positive work-family culture, as well as adverse outcomes of high levels of work-family conflict for organisations, employees and their families (Butts et al., 2013; Greenhaus et al., 2012). It was apparent in this research that the availability of FSPs and work-family conflict were not related; however, FSP availability, along with other work-related variables, did predict work-family culture. It was then a work-family culture that was a strong predictor of work-family conflict. This research demonstrated the importance of providing FSPs due to the positive contribution they make to workplace culture. Research demonstrates that merely providing FSPs is not enough as employees often do not utilise them as they fear organisational resentment or negative career consequences (Wayne & Casper, 2016). It should be noted that as FSPs are not legally required, few companies, in the absence of a considered approach such as a business case, provide all the policies a workforce often requires (Strachan, French & Burgess, 2010).

However, some companies are addressing this change and the needs of their employees by providing FSPs such as childcare assistance, work flexibility, support of eldercare and well-being programmes. Some good examples can be found. For instance, the global engineering company ARUP publicly states that their approach to work encompasses "…fair dealings with staff and clients" and their health and safety policy, as well as their diversity and inclusion policy, incorporates many aspects which support employee work-life balance (ARUP, 2016). Another example from the construction sector is Probuild, a major Australian construction company. They have many work arrangements aimed at promoting work-life balance. These include:

- flexible work arrangements: including part-time work hours, working from home, job-share, flexible start/finish times and *ad hoc* informal flexibility;
- well-being initiatives: including 'flu shots, health checks, exercise groups and activities, hearing tests, healthy lifestyle news/seminars, shower facilities, financial planning advice, personal counselling services, salary sacrifice and community relations projects;
- career development: including apprentice and graduate programmes, further education assistance, performance management, training and development aligned to performance management, and professional association/industry memberships;
- leave arrangements: including maternity leave, paternity leave, purchased leave, personal leave (sick and carers), compassionate leave, Probuild productivity leisure days, long service leave, jury service, and time for defence force reservist and emergency services volunteer activities (McMahon & Pocock, 2011: 5).

These initiatives demonstrate many companies' willingness to address the issue of work-life conflict. The public nature of the disclosure illustrates the value they have regarding future staff recruitment and corporate culture promotion. When

FSPs are considered, to maximise their value they must be developed in conjunction with all staff so that every sector of the workforce benefits equally. Making incorrect, but often well meant, assumptions about employees' needs may result in the wrong FSPs being implemented and not utilised.

FSPs and good management practices which impact on work-family culture are important for engineers who are also parents, particularly those who are mothers. It was apparent there was a general reticence by the men in the sample to make use of the available FSPs. It would appear the male breadwinner model is alive and well in engineering and women have to work around this and the practices it embeds in the work-life culture of the company. It is evident that mothers are already modifying their work to suit family responsibilities, but this may not be at the level they may personally prefer. Men making less use of FSPs are not uncommon, and lower uptake of FSPs by both men and women in science, technology, engineering and mathematics (STEM) fields has been found (Lundquist et al., 2012). It appears the male-dominated nature of groups controls the use of FSPs by both men and women, with masculine cultures tending to drive mothers out of organisations (Cahusac & Kanji, 2014).

The work-related variables that influenced work-family culture were collegial support, role conflict and work hours. While previous studies have linked supervisory support to work-family culture, this study demonstrated the importance of collegial support. Support from colleagues can be considered as an additional resource and one which contributes to organisational support (Allen et al., 2013). Organisations should, therefore, encourage this support to develop, which may be achieved through developing a better understanding of one's work mates. Fostering social interaction, both within and outside the organisation, can assist as well as a provision of teamwork skills' training and interpersonal skills' development (Aguinis & Kraiger, 2009). Flatter organisational structures are also considered useful in developing stronger ties between colleagues as it reduces an 'us' and 'them' culture which emanates from hierarchical structures (Chiaburu & Harrison, 2008).

Role conflict can occur when there is ambiguity regarding what a job involves as well as what the organisation requires. While all roles will have some role ambiguity and may have creativity and learning advantages, when job expectations and performance goals are unclear stress and dissatisfaction can develop (Savelsbergh et al., 2012). This uncertainty or lack of job clarity can be a result of poor management skills such as unclear communications, imprecise directives, and contradictory instructions.

This uncertainty can lead to poor performance as well as distress for employees. Having open communication, clear goals and a mutual understanding of one's role within the larger context can assist, along with the establishment of responsibilities, time frames and regular feedback (Hassan, 2013). Davis (2015) highlights the importance of both scheduled and unscheduled supervision to role clarity. Relying on unscheduled supervision alone can be problematic as it is typically brief and occurs to meet a specific, and often urgent, situation. While a necessary complement, regularly scheduled meetings should be adopted to address the breadth and depth of roles, providing reliable and useful discussion (Davis, 2015). The correct 'on-boarding' of newcomers is particularly important as they can experience greater levels of uncertainty regarding roles and expectations. Lapointe et al. (2014) suggest formal socialisation tactics are useful in building confidence in these situations.

Finally, a reduction of work hours also needs to be considered, especially since it is closely linked with both work-family culture and work-family conflict. It is now also well understood that long working hours are detrimental to both physical and mental health and associated with poorer work performance (Sparks et al., 2013; Caruso, 2014). For instance, it has been demonstrated that sleep hours are reduced and sleep disturbances increased when employees' weekly work hours are long (Sasaki et al., 1999). With sleep deprivation comes a decline in cognitive functioning, ability to concentrate, motor skills, memory and ability to learn new information as well as higher rates of errors and fatigue-related injuries (Goel et al., 2009). In addition, long work hours are associated with a higher prevalence of depression, anxiety, fatigue and poorer mood, recovery and perceived health (Kleppa et al., 2008; Caruso et al., 2004; Ettner & Grzywacz, 2001; Van der Hulst et al., 2006; Siu & Ian, 1996).

As Turner et al. (2008) found, project-orientated organisations do not deal well with matters relating to employee well-being, which includes issues relating to work-life balance. Effective HR practices and greater managerial effort and support can certainly assist. Regarding diversity, it may be that when men adopt more FSPs, overall change will occur, not only to the retention rates of female engineers but also in the type of engineering entrants and the number of women considering engineering as a viable career.

Positive change has started, and forward thinking companies are adapting to these new models of work. Perhaps the final word should go to David Cruickshanks-Boyd, National President of Engineers Australia in 2015 and Regional Director of Parsons Brinckerhoff, whose own work arrangements and views on the future of work are refreshing and may allay the fears of companies yet to embark on this much-needed change:

> My flexible work arrangement of a 9-day fortnight is terrific. I feel I am delivering for Parsons Brinckerhoff and our clients but, I get an excellent lift in well-being by having a frequent three-day weekend. I envision a workplace in the future which has greater diversity at all leadership levels, enabled by fully implemented flexible working with obvious benefits to organisational performance. In parallel, I envisage home environments in which all members can contribute equally to the nurture of their children, the care of elderly loved ones, the contribution to their communities and the pursuit of endeavours outside of work. Adopting such a workplace culture delivers clear organisational benefits.
>
> (Engineers Australia, 2015: 2)

References

Aguinis, H. & Kraiger, K. (2009), Benefits of training and development for individuals and teams, organizations, and society, *Annual Review of Psychology*, Vol. 60, pp. 451–474.

Allen, T. D. (2001), Family-supportive work environments: The role of organizational perceptions, *Journal of Vocational Behavior*, Vol. 58(3), pp. 414–435.

Allen, T. D. (2012), The work and family interface. In: S. W. J. Kozlowski (ed.), *The Oxford Handbook of Organizational Psychology*, Oxford University Press, New York, NY, pp. 1163–1198.

Allen, T. D., Johnson, R. C., Kiburz, K. M. & Shockley, K. M. (2013), Work-family conflict and flexible work arrangements: Deconstructing flexibility, *Personnel Psychology*, Vol. 66(2), pp. 345–376.

Andreassi, J. K. & Thompson, C. A. (2008), Work-family culture: Current research and future directions. In: K. Korabik, D. S. Lero & D. L. Whitehead (eds.), *Handbook of work-family integration: Research, theory, and best practices*, Academic Press, Amsterdam, pp. 331–351.

ARUP. (2016), *Policies*. Available at: www.arup.com/about_us/a_people_business/policies. Date of access: 17 October 2016.

Australian Bureau of Statistics. (2009), *Australian social trends*, September 2009 (Cat. No. 4102.0), Australian Bureau of Statistics, Canberra.

Bloom, N., Kretschmer, T. & Van Reenen, J. (2011), Are family-friendly workplace practices a valuable firm resource? *Strategic Management Journal*, Vol. 32(4), pp. 343–367.

Browne, R. (2010), Life balances still not working. *Sydney Morning Herald*. Available at: www.smh.com.au/national/life-balance-still-not-working-20100731-110pg.html Date of access: 10 September 2016.

Budd, J. W. & Mumford, K. A. (2006), Family-friendly work practices in Britain: Availability and perceived accessibility, *Human Resource Management*, Vol. 45(1), pp. 23–42.

Butts, M. M., Casper, W. J. & Yang, T. S. (2013), How important are work–family support policies? A meta-analytic investigation of their effects on employee outcomes, *Journal of Applied Psychology*, Vol. 98(1), p. 1.

Byron, K. (2005), A meta-analytic review of work–family conflict and its antecedents, *Journal of Vocational Behavior*, Vol. 67(2), pp. 169–198.

Cahusac, E. & Kanji, S. (2014), Giving up: How gendered organizational cultures push mothers out, *Gender, Work & Organization*, Vol. 21(1), pp.57–70.

Caruso, C. C. (2014), Negative impacts of shiftwork and long work hours, *Rehabilitation Nursing*, Vol. 39(1), pp. 16–25.

Caruso, C.C., Hitchcock, E.M., Dick, R.B., Russo, J.M. & Schmit, J.M. (2004), Overtime and extended work shifts: Recent findings on illnesses, injuries, and health behaviors. Department of Health and Human Services, Public Health Service, Centers for Disease Control and Prevention, National Institute for Occupational Safety and Health, Cincinnati, OH. *DHHS (NIOSH) Publication No. 2004-143*.

Chiaburu, D. S. & Harrison, D. A. (2008), Do co-workers make the place? Conceptual synthesis and meta-analysis of lateral social influences in organizations, *Journal of Applied Psychology*, Vol. 93(5), pp. 1082–1103.

Cohen, J.W. (1988), *Statistical power analysis for the behavioural sciences* (2nd edn). Erlbaum, New York, NY.

Davis, J. K. (2015), Supervision of peer specialists in community mental health centers: Practices that predict role clarity, *Social Work in Mental Health*, Vol. 13(2), pp. 145–158.

De Sivatte, I., Gordon, J. R., Rojo, P. & Olmos, R. (2015), The impact of work-life culture on organizational productivity, *Personnel Review*, Vol. 44(6), pp. 883–905.

Eby, L. T., Casper, W. J., Lockwood, A., Bordeaux, C. & Brinley, A. (2005), Work and family research in IO/OB: Content analysis and review of the literature (1980–2002), *Journal of Vocational Behavior*, Vol. 66(1), pp.124–197.

Edwards, J. R. & Rothbard, N. P. (2000), Mechanisms linking work and family: Clarifying the relationship between work and family constructs, *Academy of Management Review*, Vol. 25(1), pp.178–199.

Engineering UK. (2016), *Engineering UK 2016: Synopsis, recommendations and calls for action*, Engineering UK, London.

Engineers Australia. (2015), *Women in engineering creating effective workplaces for now and in the future – Flexible workplace strategies*. Available at: www.engineersaustralia.org.au/ sites/default/files/shado/Learned%20Groups/ Interest%20Groups/Women%20in%20 Engineering/wie_industry_blueprint_web_2015.pdf Date of access: 19 October 2016.

Eisenberger, R., Huntington, R., Hutchison, S. & Sowa, D. (1986), Perceived organizational support, *Journal of Applied Psychology*, Vol. 71(3), pp. 500–507.

Ettner, S. L. & Grzywacz, J. G. (2001), Workers' perceptions of how jobs affect health: A social ecological perspective, *Journal of Occupational Health Psychology*, Vol. 6(2), p. 101.

Faulkner, W. (2009), Doing gender in engineering workplace cultures. I. Observations from the field, *Engineering Studies*, Vol. 1(1), pp. 3–18.

Goel, N., Rao, H., Durmer, J. S. & Dinges, D. F. (2009), Neurocognitive consequences of sleep deprivation, *Seminars in Neurology*, Vol. 29(4), pp. 320–339.

Goode, W. J. (1960), A theory of role strain, *American Sociological Review*, Vol. 25(4), pp.483–496.

Gray, M. & Tudball, J. (2003), Family-friendly work practices: Differences within and between workplaces, *The Journal of Industrial Relations*, Vol. 45(3), pp. 269–291.

Greenhaus, J. H. & Beutell, N. J. (1985), Sources of conflict between work and family roles, *Academy of Management Review*, Vol. 10, pp. 76–88.

Greenhaus, J. H., Ziegert, J. C. & Allen, T. D. (2012), When family-supportive supervision matters: Relations between multiple sources of support and work–family balance, *Journal of Vocational Behavior*, Vol. 80(2), pp. 266–275.

Hackman, J. R. & Oldham, G. R. (1975), Development of the Job Diagnostic Survey, *Journal of Applied Psychology*, Vol. 60, pp.159–170.

Hassan, S. (2013), The importance of role clarification in workgroups: Effects on perceived role clarity, work satisfaction, and turnover rates, *Public Administration Review*, Vol. 73(5), pp.716–725.

Hill, E., Hawkins, A., Ferris, M. & Weitzman, M. (2001), Finding an extra day a week: The positive influence of perceived job flexibility on work and family life balance, *Family Relations*, Vol. 50, pp. 49–58.

Jones, F., Burke, R. J. & Westman, M. (2013), *Work-life balance: A psychological perspective*, Psychology Press, New York.

Kahn, R. L., Wolfe, D. M., Quinn, R. P., Snoek, J. D. & Rosenthal, R. A. (1964), *Organizational stress: Studies in role conflict and ambiguity*, John Wiley, Oxford, England.

Kaspura, A. (2015), *The engineering profession: A statistical overview* (11th edn), Institution of Engineers Australia, Barton ACT.

Keith, T.Z. (2006), *Multiple regression and beyond*, Pearson, Austin, TX.

Kleppa, E., Sanne, B. & Tell, G. S. (2008), Working overtime is associated with anxiety and depression: The Hordaland Health Study, *Journal of Occupational and Environmental Medicine*, Vol. 50(6), pp. 658–666.

Lambert, V. A. & Lambert, C. E. (2001), Literature review of role stress/strain on nurses: An international perspective, *Nursing & Health Sciences*, Vol. 3(3), pp. 161–172.

Lapierre, L. M. & Allen, T. D. (2012), Control at work, control at home, and planning behavior implications for work–family conflict, *Journal of Management*, Vol. 38(5), pp. 1500–1516.

Lapointe, É., Vandenberghe, C. & Boudrias, J. S. (2014), Organizational socialization tactics and newcomer adjustment: The mediating role of role clarity and affect-based trust relationships, *Journal of Occupational and Organizational Psychology*, Vol. 87(3), pp.599–624.

Leedy, P.D. & Ormrod, J.E. (2016), *Practical research: Planning and design*, (11th edn), Pearson, London.

Lewis, R. A. & Sussman, M. B. (2013), *Men's changing roles in the family*, Routledge, New York.

Lewis, S. (2001), Restructuring workplace cultures: The ultimate work-family challenge? *Women in Management Review*, Vol. 16(1), pp. 21–29.

Lingard, H. & Francis, V. (2009), *Managing work-life balance in construction*, Routledge, Abingdon.

Lingard, H. C., Francis, V. & Turner, M. (2010), The rhythms of project life: A longitudinal analysis of work hours and work–life experiences in construction, *Construction Management and Economics*, Vol. 28(10), pp. 1085–1098.

Lundquist, J. H., Misra, J. & O'Meara, K. (2012) Parental leave usage by fathers and mothers at an American university. *Fathering*, Vol. 10(3), pp. 337–363.

McMahon, C. & Pocock, B. (2011), *Doing things differently: Case studies of work-life innovation in six Australian workplaces*, Centre for Work + Life University of South Australia, Adelaide.

Michel, J. S., Kotrba, L. M., Mitchelson, J. K., Clark, M. A. & Baltes, B. B. (2011), Antecedents of work–family conflict: A meta-analytic review, *Journal of Organizational Behavior*, Vol. 32(5), pp. 689–725.

Netemeyer, R. G., Boles, J. S. & McMurrian, R. (1996), Development and validation of the work-family conflict and family-work conflict scales, *Journal of Applied Psychology*, Vol. 81, pp. 400–410.

Ngo, H. Y., Foley, S. & Loi, R. (2009), Family friendly work practices, organizational climate, and firm performance: A study of multinational corporations in Hong Kong, *Journal of Organizational Behavior*, Vol. 30(5), pp. 665–680.

Page, A., Baird, M., Heron, A. & Whelan, J. (2009), *Taking care: Mature age workers with elder care responsibilities*. Women and Work Research Group, University of Sydney, Sydney.

Pallant, J. (2011), *SPSS survival manual*, (4th edn), Allen and Unwin, Crow's Nest.

Peers, S. (2016), *Statistics on women in engineering*, Women's Engineering Society, London.

Rizzo, J. R., House, R. J. & Lirtzman, S. I. (1970), Role conflict and ambiguity in complex organizations, *Administrative Science Quarterly*, pp. 150–163.

Rofcanin, Y., Las Heras, M. & Bakker, A. B. (2016), Family supportive supervisor behaviors and organizational culture: Effects on work engagement and performance, *Journal of Occupational Health Psychology*, Vol. 21(3), pp. 296–308.

Sasaki, T., Iwasaki, K., Oka, T., Hisanaga, N., Ueda, T., Takada, Y. & Fujiki, Y. (1999), Effect of working hours on cardiovascular-autonomic nervous functions in engineers in an electronics manufacturing company, *Industrial Health*, Vol. 37(1), pp. 55–61.

Savelsbergh, C., Gevers, J. M., Van der Heijden, B. I. & Poell, R. F. (2012), Team role stress: Relationships with team learning and performance in project teams, *Group & Organization Management*, Vol. 37(1), pp. 67–100.

Schmidt, S., Roesler, U., Kusserow, T. & Rau, R. (2014), Uncertainty in the workplace: Examining role ambiguity and role conflict, and their link to depression – a meta-analysis, *European Journal of Work and Organizational Psychology*, Vol. 23(1), pp. 91–106.

Siu, O. L. & Ian, D. (1996), Psycho-social factors at work and workers' health in Hong Kong: An explanatory study. *Bulletin of the Hong Kong Psychological Society*, Vol. 34/35, pp. 30–56.

Skinner, N. & Pocock, B. (2014), *The persistent challenge: Living, working and caring in Australia in 2014. The Australian Work and Life Index.* Centre for Work + Life, University of South Australia, Adelaide.

Soltani, I., Hajatpour, S., Khorram, J. & Nejati, M. (2013), Investigating the effect of role conflict and role ambiguity on employees' job stress: Articulating the role of work-family conflict, *Management Science Letters*, Vol. 3(7), pp. 1927–1936.

Sparks, K., Cooper, C. L., Fried, Y. & Shirom, A. (2013), The effects of working hours on health: A meta-analytic review. In: C. Cooper (ed.), *From stress to wellbeing Volume 1*, Palgrave Macmillan, London, UK, pp. 292–314.

Strachan, G., French, E. L. & Burgess, J. (2010), *Managing diversity in Australia: Theory and practice*, McGraw-Hill, New York.

Thompson, C. A., Beauvais, L. L. & Lyness, K. S. (1999), When work–family benefits are not enough: The influence of work–family culture on benefit utilization, organizational attachment, and work–family conflict, *Journal of Vocational Behavior*, Vol. 54(3), pp. 392–415.

Turner, R., Huemann, M. & Keegan, A. (2008), Human resource management in the project-oriented organization: Employee well-being and ethical treatment, *International Journal of Project Management*, Vol. 26(5), pp. 577–585.

Van der Hulst, M., Van Veldhoven, M. & Beckers, D. (2006), Overtime and need for recovery in relation to job demands and job control, *Journal of Occupational Health*, Vol. 48(1), pp.11–19.

Watts, J. H. (2009), 'Allowed into a man's world' meanings of work-life balance: Perspectives of women civil engineers as 'minority' workers in construction, *Gender, Work & Organization*, Vol. 16(1), pp. 37–57.

Wayne, J. H. & Casper, W. J. (2016), Why having a family-supportive culture, not just policies, matters to male and female job seekers: An examination of work-family conflict, values, and self-interest, *Sex Roles*, Vol. 75(9), pp. 459–475.

Wilkinson, S. (2006), Women in civil engineering. In: A.W. Gale & M. Davidson (eds), *Managing diversity and equality in construction: Initiatives and practice*, Taylor & Francis Group, London.

Part III
Thinking on new practices

11 Developing health and safety competence for people in construction

David Oloke and Philip McAleenan

But only he who, himself enlightened, is not afraid of shadows

(Immanuel Kant)

Background

Construction is a people-driven activity/process, which contributes immensely to the development of society for the benefit of humankind (Oloke et al., 2007). However, the inherent nature of people makes errors, fatigue, mistakes, incidents, and, at times, accidents inevitable (Manase et al., 2011). This makes management of health and safety an essential process of people management during the construction process. Several parties within a construction project contribute to ensuring the overall health and safety of people. These parties include mainly clients, designers, contractors, safety consultants, and workers.

The aim of this chapter is to offer a cursory glimpse into the requirements for the various parties in a construction project to perform their duties within the context of the project. Each party has a goal to deliver on the project, from the client, who has the responsibility of defining the project and its scope, to the designer, who shapes the requirements, to the contractor or construction team, who is responsible for implementing the set goals. Construction safety risks are thus borne mostly at the construction site or off-site production centre, and, as such, these need to be reduced as much as possible before works get to site (HSE, 2015). Also, all residual risks need to be well mitigated or managed on-site to minimise their negative impact on the project. This is why competence is an essential requirement for everyone involved.

In achieving the above aim, however, this chapter elaborates on the following.

- Competence is defined as it relates to all parties concerned, and in the context of health and safety management; the place of knowledge, skills, and experience in the development of a competent construction workforce is also explored.

- The general procurement framework for all parties involved in the construction contracts is examined within a competency framework.
- Development of a suitable framework that seeks to address possible measures that will facilitate the development of the most relevant health and safety competence in the workforce.

The chapter concludes with recommendations that help guide the reader on how to implement the framework.

Duties and roles of participants in a construction project

Each construction project has people involvement; as various parties contribute to the development of the project. While no two construction projects are ever the same, typically any project would have at least two of the following categories of participants:

- clients;
- designers;
- contractors;
- safety consultants/professionals; and
- workers.

The above parties are usually governed in respect of their health and safety obligations, as may be stipulated by the specific country legislation that would apply to them. In the UK, for instance, this would be the Construction Design and Management (CDM) Regulations – the most current edition being the CDM 2015 in Great Britain, and the CDM 2016 in Northern Ireland (CDM, 2015). According to McAleenan and Oloke (2015), the following is an overview of these requirements.

Clients are organisations or individuals for whom a construction project is carried out. The customer is required to make suitable arrangements for managing a project (CONIAC, 2015a). This includes making sure:

- that other duty holders are appointed;
- that sufficient time and resources are allocated;
- that relevant information is prepared and provided to other duty holders;
- that the principal designer and the principal contractor carry out their duties; and
- that welfare facilities are provided.

Designers are people or organisations who prepare or modify designs for a building, product, or system relating to construction work as part of their business. These would normally be architects and engineering designers. However, experience has shown that some clients and other consultants, e.g. cost consultants, may also undertake designer duties in the context of the definition above

(CONIAC, 2015d). They are required to eliminate, reduce, or control foreseeable risks that may arise during construction, and when maintaining and using a building once it is built. It is expected that other members of the project team help them fulfil their duties, however (Oloke, 2015).

Contractors are the parties that do the actual construction work, and they can be either individuals or companies. They plan, manage, and monitor construction work under their control so that it is carried out without risks to health and safety. Where projects involve more than one contractor, the contractors are expected to coordinate their activities with others in the project team. In particular, they are projected to comply with directives given to them by other significant parties, e.g. contractors have to comply with instructions given by the principal contractor. They are also responsible for preparing a construction-phase plan for single-contractor projects. This plan is to contain the risk-assessment log and the method statements that ensure that the residual risks are appropriately mitigated as far as reasonably practicable (CONIAC, 2015c).

Safety consultants or professionals are those with specialist duties of oversight of health and safety on construction projects. Under the UK CDM Regulations, although every duty holder has a duty of care for their work and actions, the regulations designate a role called the principal designer, who is a designer appointed by the client in a project involving more than one contractor, to ensure this. They can be an organisation or an individual with sufficient knowledge, experience, and ability to carry out the role. Their work is to plan, manage, monitor, and coordinate health and safety in the pre-construction phase of a project. This includes identifying, eliminating, or controlling foreseeable risks, and ensuring that designers carry out their duties. They are also to prepare and provide relevant information to other duty holders and particularly to the contractor (or principal contractor), to help them plan, manage, monitor, and coordinate health and safety in the construction phase (CONIAC, 2015b).

A **worker** is an individual working for or under the control of contractors on a construction site. Workers, as well as their employers, have duties, as people who are working for or under the control of contractors on a construction site (CONIAC, 2015e).

It is expected that workers are consulted about matters that affect their health, safety, and welfare and that they take care of their health and safety, and that of others who may be affected by their actions. Workers are also to report anything that they see that is likely to endanger either their own or others' health and safety and to cooperate with their employer, fellow workers, contractors, and other duty holders (ibid.).

Definition of competence

It has been recognised worldwide since 2003 that competence in health and safety should be an integral aspect of vocational and technical education. The Québec City Protocol (ISSA 2003: 6) adopted the principle that "OHS [occupational health and safety] competencies associated with each step in the performance of a task must be integrated into the educational process for a given occupation", and that "[m]astery of the required knowledge and recommended practices relating to OHS should be a focus of evaluation integrated into the educational process". In advocating partnerships between the education community and those responsible for prevention, the Protocol called for competencies (knowledge and skills) that make it possible for professional, vocational and technical workers to

- adopt safe work methods and techniques;
- detect potential hazards, evaluate risks and implement prevention methods to eliminate these dangers or to control those that cannot be removed;
- adopt appropriate behaviour given the risks involved in work situations;
- participate in implementing different prevention strategies; and
- enable employers and workers to exercise their rights and responsibilities (ISSA, 2003: 8).

In 2011, informed by the Protocol, the Institution of Civil Engineers (ICE) in Northern Ireland collaborated with industry, academia, and the Health and Safety Executive for Northern Ireland, to agree to a set of objectives for OHS in construction (McAleenan et al., 2011: 6–7) and in 2012 launched the Construction Industry (NI Region) Safety and Health Declaration, which committed signatories to "integrate safety and health competency development into further and higher education processes". To date, undergraduate and post-graduate courses have developed new and improved OHS modules within construction sector degrees, and at Ulster University a safety engineering and disaster management course was introduced in 2016.

In several respects, the above measure in the Protocol is simply a list of objectives, or outcomes of competency, in practice, where it is left to education bodies and industry to establish the specific stipulations for particular sectors, professions, and vocations. Given the nature of the Protocol, further elaboration may have created restrictions on how implementation bodies interpret and apply the development of OHS competencies in various industries, and it is worth exploring here what may be added to identify what competencies are required by different sectors of the construction sector.

Hazard identification and evaluation, and control of the work activity to ensure prevention, requires that the competent person has thorough knowledge of the hazards applicable to their job, and that they be knowledgeable of and able to apply a hazard control matrix (for example, the Management of Health and Safety at Work Regulations 1999: Schedule 1: General Principles of Prevention, in Great Britain) appropriately. Additionally, it is important that all persons can manage hazards within their particular area of work and sphere of competence. It is not appropriate that this task falls to those with the title of "manager"; the skilled worker, as part of a competent team, is capable of identifying, establishing and maintaining control of their work activities and the environment in which they occur. This means their activities and taking appropriate decisions about what needs to be done, including decisions about whether it is safe to proceed or not. Those on site tasked with the role of management in the formal sense are not managing the activity; they are competent to manage the resources, materials, finances, time, etc., and to ensure that the workers carrying out the activity have sufficient resources to establish control and carry out the activity safely.

This raises the issue of the skills and qualifications needed to perform tasks on construction sites. Mertens (1999) developed a tripartite classification of competencies applicable to all workers (technical, vocational, and professional), namely basic competencies, generic competencies, and specific competencies. Basic qualifications/competencies are those that facilitate an ability to continue learning and adapt to an organisation's requirements. There are common elements to basic competencies that are transferable across industries and activities within industries, such as academic qualifications, e.g. the ability to read and write, personal development skills, e.g. level of maturity, ability to relate to others, etc., and teamwork skills, e.g. the ability to work and cooperate with others and participate in the setting and achievement of common objectives.

Generic competencies have been defined by Boyatzis (in Mertens 1999) as "the underlying characteristics of an individual which maintain a causal relationship with effective or superior performance in the job". The ability to carry out a work activity to achieve specific results through specific actions within a given context is regarded as central to effective performance, and it covers a range of standards of outcomes, from effective, i.e. that which is required to meet the objectives of the task, to superior, i.e. that which is of a higher standard, or that which demonstrates advanced skills and abilities.

Specific competencies relate to special services or products, or products or services that are specific to an organisation. Mertens (1999) states that specific competence is "a holistic or integrated relationship, incorporating the complex combination of attributes (knowledge, attitudes, values and abilities) needed for performance in specific situations". People with this type of competence could include electricians who are specialists in high-voltage transmission lines or in equipment wiring, and engineers who are specialists in road construction or bridge building.

Awareness of and ability to contribute to safety, health, and welfare is integral to all three types of competence. As an individual's competence increases, so does their ability to perform in a safe and healthful manner. This is recognised by professional bodies, which often require safety competencies as part of the membership route to chartership and continuing professional development (CPD). The ICE, for example, requires that all of its members have a "sound knowledge of health and safety" in the area of engineering, and those who can demonstrate that they have the required level of health and safety for all areas of construction may enter onto the Health and Safety Register if they are newly qualified as a professional engineer or other construction professional (Level 1), or if they have a lead role in civil engineering design, and experience of construction processes that particularly relate to health and safety (Advanced level) (ICE, 2016).

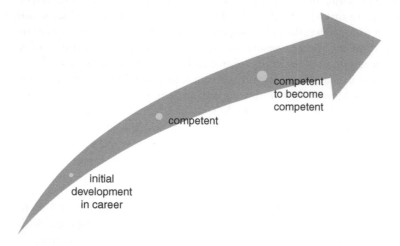

Figure 11.1 The development of competence (Source: McAleenan & McAleenan, 2009)

The definitions of competence are often limited to references to knowledge (including qualifications), skills, and experience. This is a common perspective, but if we explore competence as a concept, it may reasonably be stated that competence development represents a unique journey from dependence to independence, from a position where teachers, mentors, and supervisors are important players in ensuring that the learner can practise and carry out tasks safely, making errors and developing skills, to a position where the competent person is capable of accepting, understanding and acting upon information received, and doing so in a manner that ensures safety and quality of the product or the service. In educational terms, knowledge must lead to understanding and understanding to an ability to exercise control. In the context of competence, in construction, this means control of the activities and processes on site. Control emerges from an agency, or the autonomous capacity to make decisions, and this, in turn, occurs effectively when the competent person can develop solutions from their understanding of the situation, from their ability to see continuity (Dawkins, 1993) within processes and activities. Thus, at the point of excellence, the competent person is capable of becoming competent in different and broader areas. In other words, they are competent to become competent and can continuously extend the boundaries of their knowledge and abilities.

Procurement framework and health and safety risk management

Construction often involves a complex set of operations that lead to the desired output of a new building/facility, or upgrading/refurbishment of the same. Several different human and environmental factors that occur throughout the process often generate several independent and interdependent hazards (Oloke et al., 2007).

Procurement in construction is the process that facilitates the purchase or development of buildings or facilities. The design and construction of a building can be procured through various routes. Procurement routes normally align with a strategy that fits the long-term objectives of the client's business plan. Clients would normally consider speed, cost, quality, long-term ownership, various risks, financing, and other project constraints or limitations in arriving at a decision on which procurement route to take. According to the joint contributions of major professional bodies to the Designing Wiki (Designing Buildings Wiki, 2016) platform, common procurement routes at the moment include (but are not limited to):

- traditional contracts;
- design-and-build contracts (single-stage, or two-stage);
- construction management contracts; and
- others, such as private finance initiatives (PFIs), framework agreements, etc.

The client and due diligence

Before examining the specific challenges relating to health and safety management under the different types of contract, it is worth looking at the due diligence obligations of the client in appointing consultants and contractors. Regardless of the type of contract, the client, whether experienced or inexperienced, must ensure that design consultants, contractors, and others in the supply chain meet the competency requirements and that they have the experience and good record necessary to bring the project to a satisfactory and safe conclusion. This means that the client must take reasonable care to ensure that they are engaging designers and contractors who have a proven track record on previous projects in which they have been engaged. The experienced client may well have internal systems by which they pre-qualify and maintain lists of approved competent persons and companies. Due diligence investigations extend to satisfying oneself that the consultants and the contractors have the relevant capacity (material, financial and other resources) to undertake the project.

The inexperienced client, usually in the private sector, and including domestic clients, may not have the competencies in-house to conduct a due diligence investigation and may need to have recourse to a third party to assist them in making their decision about who to appoint or to designate to build (Bulbul, 2011). In essence, inexperienced clients require the services of an independent client advisor. The professions are a reliable source of assistance in this respect, and clients may approach the appropriate professional body for independent advice and information about the appropriateness of that particular profession, and to solicit a list of local members who can conduct a due diligence investigation. In the UK and Northern Ireland, the CDM 2015 and the CDM (NI) 2016, respectively, require designers (and contractors) to "have the skills, knowledge and experience necessary to fulfil the role and complete the tasks they are appointed to undertake in a manner that secures the health and safety of any person affected by the project", and they require companies to have the organisational capacity to complete required tasks. This means that professional institutions:

- must know all the health and safety issues relevant to the industry and the work of their members;
- must ensure that appropriate health and safety instruction is included in undergraduate, postgraduate and professional development programmes and courses of study;
- must ensure that all members/candidates can demonstrate prior to and continuously throughout their membership an up-to-date appropriate level of knowledge and skills in health and safety; and
- must ensure that member, including organisations that are members, are providing services within their areas of expertise and competence.

In this way, the independent or inexperienced client can be reassured of the competence of those recommended by the institution.

As we examine the different types of contract below, the challenges related to health and safety management, and the competency requirements detailed at the end of the sections about each type of contract, may apply to other types of contract, even though they are not detailed again, so as to avoid duplication. The respective challenges and competency requirements are to be read with that proviso.

Contracts

The **"traditional" procurement** route is the most commonly used method of procuring building works. Professionals, particularly contractors, sometimes refer to the process as "design-bid-build", or simply as "bid-build" (Designing Buildings Wiki, 2016).

The process entails the client first appointing consultants (architects, engineers, and others) to design the project in detail, and then to prepare tender documentation. The documentation should normally include drawings, work schedules, and bills of quantities. Contractors are then invited to submit tenders for the construction of the project, usually on a single-stage, competitive basis. Such a contract can be referred to as a "traditional contract". The contractor is thus not responsible for the design, other than temporary works, although some traditional contracts do provide for the contractor to design specific parts of the works.

The client normally retains the design consultants during the construction phase, to assist in preparing any required additional design information, to review any design proposals by the contractor, and also to inspect the works. The architect (mostly in building projects) or another consultant may also be appointed to administer the contract.

Depending on the arrangements for financing and operating the contracts, traditional construction contracts are most commonly lump-sum contracts (in which case a budget is secured, and rates are worked in). However, measurement contracts and cost reimbursement contracts can also be used for "traditional" projects, where design and construction are separate, sequential activities, due to the nature of the project. In the latter case, work is measured as progress is made, and the appropriate remunerations are made to the contractor or the consultants, as the case may be.

This form of procurement can be used for all clients, regardless of their level of experience. It is considered to be a low-risk method of contracting for the client, as the contractor takes the financial risk for construction. If the design is fully developed before tender, the client has greater certainty about design quality and cost. However, this method of procurement can be slower than other forms of contracting, because the contractor can be appointed only after the design has been completed. Also, improving buildability and packaging of proposals as they develop can be very difficult.

However, as in some cases, if design information is incomplete at tender stage, or significant variations are required after the contractor has been appointed, the cost to the client can be significant. This is thus a downside of traditional procurement.

In this model, the lead designer and other design team members are appointed at the earliest stages of the project. In line with their professional obligations to be competent on the health and safety requirements necessary for the project, it is at this stage that the design team should be considering the implications of their design on the health and safety of workers during the construction phase, the use and maintenance phases, and the refurbishment and demolition phases. In other words, they need to take into account both the immediate needs and the long-term needs and produce designs that eliminate design and construction hazards, before the project goes to tender. This may involve, for example, looking at alternative designs, or materials and processes for fabrication off-site. At this stage, it is useful to have a principal contractor involved who is familiar with hazards associated with construction, and whose competence will bring a fresh pair of eyes to any potential problems with safety during the building phase, and thus allow the designer to modify designs to avoid or reduce those problems.

When the tender documentation, including bills of quantities, etc., has been prepared and submitted to prospective tenderers, the contractors must have the requisite knowledge of the health and safety requirements of the project, and they must take account of these requirements when costing and planning for the project. This will include consideration of safer means of construction, engineering and mechanical control methods appropriate to the project, specialist training for unique or unusual aspects of the project, or any other costs associated with controlling hazards or residual risks. In particular, costing the project should include proper remuneration for a competent workforce, which each contractor is expected to engage in the project. This aspect of health and safety management is often overlooked, and low-paid unskilled or poorly skilled workers are engaged who require increased levels of training and supervision. Where a multi-lingual workforce is engaged, the costing must consider engaging managing and supervisory engineers who are the same nationality as the workforce, to ensure that communication is clear and understandable to all.

A **design-and-build contract** is a procurement route in which the main contractor is appointed to design and construct the works. Clients prefer the design-and-build route at times, as it gives a single point of responsibility for delivering the project. Projects so procured can follow either a single-stage or a two-stage tender process. Either way, responsibility for management of risk resides mostly with the contractor(s).

The challenge to health and safety with this type of contract lies like the relationship of the designer to the operations manager, specifically where they are employed by the same company. At the outset, establishment of the health and safety objective before any work commences on the design will establish how health and safety are to be managed on the project. An important consideration is to recognise and respect the autonomy and competence of the different professionals that comprise the design-and-build team and to allow that the advice given by each is used towards successful achievement of a well-built and safe project, with the emphasis on their obligation towards meeting the client's requirements. Professional integrity is central to health and safety competence, to balance these requirements with the need to construct the project safely.

A **construction management contract** is a procurement route where the works are constructed by some different trade contractors (Designing Buildings Wiki, 2016). These trade contractors are contracted to the client but managed by a construction manager. Although similar to the traditional contract in structure, a construction management contract avoids the risks associated with a fixed price, as the management contractor is reimbursed the amounts paid to the trade contractors. He further receives a percentage of the fee. In essence, the construction manager acts as an agent for the client, administering and coordinating the trade contracts. Early appointment of the construction manager in the process is recommended, so that the client can leverage their experience to improve buildability and packaging of proposals as the project progresses (Oloke, 2015).

Team-building competence is required where some different consultants and contractors are involved. This is particularly so where the participants on the team may be there for varying durations depending on their contracts; thus, the team at the beginning of the construction process will be different from the team at later stages and the final stage of the process, with the possibility that only a few members will stay on the team for the full duration of the project.

Different contractors will bring to the table their objectives and what they want from the project. Short-term contractors may not have the same commitment to the project as those who will be there for substantially longer periods, and they may not buy into the objectives of the project as fully as may be desirable. Additionally, it is important to establish how conflicts of interest are to be identified and resolved. Management contractors should thus have a range of competencies, such as team building and support, conflict resolution, and the ability to stimulate cooperation between different parties and interests.

Design-build-finance-and-operate (public-private partnership (PPP)/Private finance initiative (PFI)/design-build-operate (DBO)/build-own-operate-transfer (BOOT)). Under this arrangement, the contractor finances the project and leases it to the client for an agreed period (perhaps 30 years), after which the development reverts to the client. The procurement process involves the appointment of a single contractor (perhaps a special-purpose vehicle (SPV), with design, construction, and facilities-management expertise, as well as funding capability). The SPV group designs and builds the project, and then operates it for the agreed period, after which the development reverts to the client (Designing Buildings Wiki, 2016).

Variations include design-build-operate (DBO) and build-own-operate-transfer (BOOT). However, the most prominent form of this type of contract is the public-private partnership (PPP), where the most common variant is the private finance initiative (PFI).

The issues in these forms of contract are similar to those in the management contract, with the complication that the main parties on the team are, for the most part, employed by one company. This raises issues about agency and the ability of team members to offer autonomous advice and make autonomous decisions. It is possible that in the management of health and safety, during both the construction phase and the operation phase, a tendency towards bureaucracy may emerge, particularly as such contracts are often held by very large companies, where there are complex hierarchies and remote points of authority. Paperwork may be generated in large quantities, particularly "risk assessments", which are often wordy, complex, and unnecessary, but serve to leave a paper trail, "in case something goes wrong". What is required here is an organisational competence that bases its health and safety management on a correct understanding of health and safety operation analysis and control, together with respect for and trust in the competencies of the people employed to design, build, and manage the project. Reporting up the line is a skill involving the production of concise and accurate reports on how health and safety are being managed, and what is required from other departments to continue to meet these needs.

Recommendations for improving health and safety competence in PiC

Developing a competency framework

The International Labour Organization (ILO) OHS conventions that led to the development of the Seoul Declaration (ILO, 2008) have been concerned with the development of a prevention culture and the identification of responsibility for OHS. Consequently, the internationally agreed-on position is that OHS is a societal responsibility, with a tripartite arrangement between government and employers and employees' organisations, to ensure development and maintenance of good preventative practices in all workplaces within the jurisdiction of the nation concerned. National OHS strategies that are being developed must be integrated with national education strategies in the development of a competency framework that takes account of the Québec City Protocol (ISSA, 2003) and integrates OHS competencies into the education of young people, as well as the basic competencies necessary to ensure that all school leavers have the necessary requirements to participate fully in employment (Mertens, 1999). In this idea, the national competency framework begins at the earliest stages of education/formation and continues throughout all levels of education (see Figure 11.2). It requires cooperation with industry and professional and vocational institutions and bodies, to ensure that generic and specific competencies are clearly identified and incorporated into the framework and that resources necessary for their development and exercise are made available. The resourcing of competency must take place both in the education of competent persons and within the workplace, to facilitate the exercise of competence without neutering the ability of individuals and teams to properly carry out their work activities.

The framework for the attainment of basic competencies should focus on both the development of employability skills and those competencies necessary for the growth and development of the full person, who is growing into an autonomous being, capable of establishing positive relationships and understanding and intervening objectively in their world (Freire, 2013). Programmes such as Freire's (2013) literacy and culture programmes have combined the development of functional skills of literacy and numeracy with personal (and community) growth.

The framework requires a close working partnership between educational stakeholders and industry, to ensure that the upper levels of learning and further

Competence Levels					Education and formation
Basic					• Social/informal learning • Formal schooling
Generic	Generic	Generic Generic	Generic	Generic	• Formal further and higher education • Work experience • Professional and vocational institutions (entry level training and CPD) (There will be some overlaps between different generic competences)
Specific	Specific	Specific	Specific	Specific	• Formal higher education • Specialist Colleges and Universities • Professional and vocational institutions (CPD and specialist courses) • Company specific training • Research and Innovation by competent persons (There will be some overlaps between some generic and some specific competences)

Figure 11.2 Education and training framework

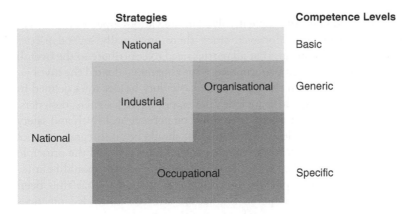

Figure 11.3 A schema for the integration of competency framework strategies

education programmes can meet the requirements of industry and society. This will entail the development of industry-specific and company-specific competency frameworks that clearly identify the skill and knowledge requirements, and that develop these within the context of social responsibility. Training and education in the relevant competencies in technical and vocational institutions should take account of the Québec City Protocol (2003), and it should fully integrate the relevant OHS requirements, not the programmes. These institutions will also act as exemplars in respect of OHS.

Specific competencies, whether technical, vocational, or professional, will be identified within the competency frameworks developed by trade bodies and professional institutions, which will identify the requirements, including specific OHS competencies, associated with both the generic and the distinct applications and specialisms within each particular trade or profession. Such bodies are ideally placed:

- to inform and guide the formal educational institutions regarding their needs, as well as to approve and accredit programmes of study;
- to develop and deliver continuing professional development courses that take into account OHS considerations;
- to specify the levels of OHS competencies necessary for full admission to membership; and
- to maintain oversight on members' continuing competence to practice.

Competency frameworks are flexible, adapting to innovation and the changing needs of society, industry, and the professions, while all the time developing individuals whose competence is respected in the workplace, through recognition of their agency, and provision of the necessary resources for them and their teams to act competently.

Summary

Managing health and safety of people involved in the various aspects of construction is an essential process. However, since there are usually several parties involved, the process also means that all parties need to contribute to the overall health and safety management performance. This chapter evaluated the involvement of the various parties in the light of the basic duty-holder roles defined in the UK CDM Regulations, which include clients, principal designers, designers, contractors, and workers. For a long time now competence in health and safety has been recognised worldwide as something that should be an integral aspect of vocational and technical education. Some countries have adopted the principle that OHS competencies linked with the performance of a task should be integrated into the educational process for a given occupation. It has thus been opined that mastery of the required knowledge and recommended practices relating to OHS should be an integral aspect of the educational process. Hazard identification and evaluation, and control of the work activity to ensure prevention requires that the competent person has a thorough knowledge of the hazards applicable to their work and that they be knowledgeable of and able to apply a hazard control matrix appropriately. Hence, an awareness of and the ability to contribute to safety, health, and welfare is essential to all aspects of competence. As an individual's competence increases, so does their ability to perform in a safe and healthful manner.

The complex set of operations that characterise construction activities relating to the establishment of new buildings/facilities, or the upgrading/refurbishment of the same, makes construction unique. Several different human and environmental factors occur throughout the process, and these often generate independent and interdependent hazards. Construction procurement facilitates the purchase or development of buildings or facilities, and procurement routes usually align with a strategy that fits the long-term objectives of the client's business plan. Organisational competence in health and safety management, and an understanding of operation analysis and control, together with respect for and trust in the competencies of people employed to design, build, and manage projects is essential. Within an organisation, line reporting is a skill involving the production of concise and accurate reports on how health and safety are being managed, and what is required from other departments to continue to meet these needs.

A framework for the attainment of core competencies is at this moment proposed. Such a framework should focus on both the development of employability skills and those competencies necessary for the growth and development of the full person involved in the construction process. This will enhance the individual's capability to establish positive relationships and engage objectively in their world. Competency frameworks must remain flexible, while all the time developing individuals whose competence is respected in the workplace, through recognition of their agency, and provision of the necessary resources for them and their teams to act competently.

References

Bulbul, T. (2011), An analysis of construction worker safety during building decommissioning and deconstruction. *Proceedings of the CIB W099 Safety and Health in Construction Conference*, August, Washington, DC.

CONIAC *see* Construction Industry Advisory Committee.

Construction Industry Advisory Committee (CONIAC) (2015a), *The Construction (Design and Management) Regulations 2015: Industry guidance for clients*. Construction Industry Training Board (CITB), King's Lynn, UK. [Online]. Available at: www.citb.co.uk/documents/cdm%20regs/2015/cdm-2015-clients-printer-friendly.pdf Date of access: 12 October 2016.

Construction Industry Advisory Committee (CONIAC) (2015b), *The Construction (Design and Management) Regulations 2015: Industry guidance for contractors*. Construction Industry Training Board (CITB), King's Lynn, UK. [Online]. Available at: www.citb.co.uk/documents/cdm%20regs/2015/cdm-2015-contractors-printer-friendly.pdf Date of access: 12 October 2016.

Construction Industry Advisory Committee (CONIAC) (2015c), *The Construction (Design and Management) Regulations 2015: Industry guidance for designers*. Construction Industry Training Board (CITB), King's Lynn, UK. [Online]. Available at: www.citb.co.uk/documents/cdm%20regs/2015/cdm-2015-designers-printer-friendly.pdf Date of access: 12 October 2016.

Construction Industry Advisory Committee (CONIAC) (2015d), *The Construction (Design and Management) Regulations 2015: Industry guidance for principal designers*. Construction Industry Training Board (CITB), King's Lynn, UK. [Online]. Available at: www.citb.co.uk/documents/cdm%20regs/2015/cdm-2015-principal-designers-printer-friendly.pdf Date of access: 12 October 2016.

Construction Industry Advisory Committee (CONIAC) (2015e), *The Construction (Design and Management) Regulations 2015: Industry guidance for workers*. Construction Industry Training Board (CITB), King's Lynn, UK. [Online]. Available at: www.citb.co.uk/documents/cdm%20regs/2015/cdm-2015-workers-printer-friendly.pdf Date of access: 12 October 2016.

Dawkins, R. (1993). Gaps in the mind, In: P. Cavalieri & P. Singer (eds), *The Great Ape Project*, St. Martin's Griffin, New York, pp. 81–87.

Designing Buildings Wiki (2016), *Procurement route*. [Online]. Available www.designingbuildings.co.uk/wiki/Procurement_route Date of access: 12 October 2016.

Freire, P. (2013 [1974]). *Education for critical consciousness*, Bloomsbury Academic, London and New York.

Great Britain (1999). *The Management of Health and Safety at Work Regulations 1999. Statutory Instrument 3242*. The Stationery Office, London.

Health and Safety Executive (HSE) (2015), *Managing health and safety in construction: Construction (Design and Management) Regulations 2015: Guidance on regulations (L153)*. HSE Books, London. [Online]. Available at: www.hse.gov.uk/pubns/priced/l153.pdf Date of access: 12 October 2016.

HSE *see* Health and Safety Executive.

ILO *see* International Labour Organization.

Institution of Civil Engineers (2016), *Specialist professional registers*. [Online]. Available at: www.ice.org.uk/careers-and-professional-development/careers-advice-for-civil-engineers/specialist-professional-registers#HSR Date of access: 10 October 2016.

International Labour Organization (ILO) (2008), *Seoul Declaration on Safety and Health at Work: The Safety and Health Summit*, ILO, Seoul.

International Social Security Association (ISSA) (2003), *Québec City protocol for the integration of occupational health and safety (OHS) competencies into vocational and technical education*. International Section on Education and Training for Prevention of the ISSA, October, Québec, Canada.

ISSA *see* International Social Security Association.

Manase, D., Heesom, D., Oloke, D., Proverbs, D., Young, C. & Luckhurst, D. (2011), A GIS analytical approach for exploiting construction health and safety information, *Journal of Information Technology in Construction*. ISSN 1874-4753. Published February at http://itcon.org/2011/21

McAleenan, C., Logan, K. & McAleenan, P. (2011), *International Workers Memorial Day, Decision Makers Conference Actions Report*. Institution of Civil Engineers, NI Region, Belfast.

McAleenan, P. & McAleenan, C. (2009), An exploration of structured and flexible approaches to recognising engineering competence. *Proceedings of CIB W099 Conference*, Melbourne.

McAleenan, C. & Oloke, D. (eds) (2015), *ICE manual of health and safety in construction*, 2nd ed., ICE Publishing, London. ISBN: 9780727760104.

Mertens, L. (1999), *Labour competence: Emergence, analytical frameworks and institutional models: With special reference to Latin America*. Montevideo, Uruguay: Cinterfor/ILO.

Oloke, D., Yu, H. & Heesom, D. (2007), Developing practitioner skills in construction health and safety management: An integrated teaching and learning approach. *Journal of Education in the Built Environment (JEBE)*, 2(1), 3–30.

Oloke, D.A.O. (2015), Responsibilities of key duty holders in construction design and management, In: C. McAleenan & D.A.O. Oloke, *ICE manual of health and safety in construction*, 2nd ed., ICE Publishing, London, pp. 15–24. ISBN: 9780727760104.

The Construction (Design and Management) Regulations 2015 (2015), [Online]. Available at: www.legislation.gov.uk/uksi/2015/51/contents/made Date of access: 12 October 2016.

12 The commodification of worker health, safety and well-being

CSR in practice

Fred Sherratt

Always recognize that human individuals are ends, and do not use them as means to your end–

(Immanuel Kant)

Background

Worker health, safety and well-being (HSW) has long been a concern; rates of worker fatalities, accidents, and ill health in construction are often significantly higher than those for other industries. For example, in the United Kingdom, the worker fatal injury rate in the construction sector is over 3.5 times the average rate across all sectors, while incidents of work-related musculoskeletal disorders and lung problems are also statistically significantly higher than in other industries (Health and Safety Executive, 2015a). These problems are repeated and magnified globally, and although precise statistics are hard to obtain and under-reporting is common, it has been estimated that around 60,000 people die on construction sites worldwide each year (Smallwood & Lingard, 2009).

The consequences of such worrying statistics are manifold. Notwithstanding fundamental ethical issues around harm and its impact on the workers themselves, the adverse effects of such damage on worker productivity and output also affect industry both at the organisational level and the wider national economy at the level of GDP, as well as placing strain on national governments as they attempt to deal with the wider social repercussions of those unable to work or who suffer from long-term health complaints, both physical and mental. As a result, worker HSW is a key focus for academic research, government intervention and legislation, and corporate investment, all seeking improvements in industry practice.

Recently, however, there has been a shift in *how* worker HSW is positioned within industry practice and government interventions: HSW is no longer a core 'stand-alone' organisational concern. Instead, it has been incorporated into the realm of corporate social responsibility or CSR (Rawlinson & Farrell, 2010). This can be seen as something of a natural progression: is CSR not an ideal home for something as fundamentally ethical as worker HSW? Is that not responsibility writ large? But such a shift should not go unchallenged. CSR also has very close relationships with other aspects of corporate governance including public relations (PR), work winning and profitability, a place where presentation and packaging are often as important, if not more so, than the content itself.

CSR also brings with it challenges of commodification (Brés & Gond, 2014). Marxist theory (1976 [1990]) has long proposed that people are themselves commodified through the sale of their labour, bringing with them a use-value, the ability of the workers to do the work, and an exchange-value in the form of their wages. Yet through this transaction the construction workers also find their individual HSW commodified; as the figures that opened this chapter indicate, worker HSW is also part of this exchange as they relinquish a part of their health and well-being (safety, or rather unsafely in the form of an accident, negating worker health) as a seemingly inevitable part of construction work. This clearly has impacts on their ability to enjoy their own HSW outside of work, but also creates an ongoing attrition of their physical abilities, impacting not only on the capacity of the worker to maintain production outputs in the future, but to enjoy future leisure as well. Within a capitalist mode of production, this is perhaps inevitable; there must be an increase in exchange value through the process, and construction workers arguably make a very personal contribution to this in the form of their HSW.

Within this context commodification does not end with the worker. It also extends to the practices of CSR itself through the creation and development of new commodities, the packaging of practices which seek to redress and mitigate this imbalance through health enhancement programmes and well-being initiatives on sites. Such activities can produce new PR 'products' that enhance and reinforce the corporate brand, increasing organisational attractiveness, and increasing the value of CSR to organisations operating within the highly competitive construction industry.

This chapter, therefore, seeks to critically explore this shift in the positioning of HSW and its rationalization under the influences of CSR and to reveal how corporate control and commodification exist within the seemingly altruistic context of HSW on construction sites.

Fundamentals of corporate social responsibility

There are ongoing debates about what CSR is. Discussants are often divided into two opposing camps broadly centred on 'stakeholder theory' or 'institutional theory' (Frynas, 2009). The former sees CSR born of stakeholder pressure, including organisational clients and customers, and their desire for some level of demonstrable corporate citizenship; the latter maintains that it simply represents another development opportunity (Pedersen, 2015; Upstill-Goddard et al., 2012) with a clearly associated business case. These two positions are not mutually exclusive, as suggested by the recognised shift from simply "doing good to do good" to "doing good to do well" (Brés & Gond, 2014).

This over-simplifies the contextual complexities of CSR. From the pure economic position, CSR has been heavily critiqued as a "misguided virtue" (Henderson, 2001), something that challenges the fundamental goal of all organisations to create profit and returns to shareholders, while operating within a wider economic context which also demands constant growth for survival. More directly it has been queried whether using shareholder resources in ways other than seeking profitable return is even a responsible activity (Vogel, 2005). Indeed, the most famous critic of CSR remains Milton Friedman (1970) who stated that:

> There is one and only one social responsibility of business – to use its resources and engage in activities designed to increase its profits so long as it stays within the rules of the game, which is to say, engages in open and free competition without deception or fraud.

There are arguably two important phrases within this oft-cited statement: 'free competition' and 'the rules of the game'. The latter will be examined later through a construction-industry specific lens. However, what must be clearly noted is that no organisation currently operates in true free competition; despite the lip-service paid to free markets by the contemporary neo-liberal doctrine, the *realities* of the neo-liberal marketplace are far different, undermined as it is by both government intervention and legislation (Ennals, 2011), making Friedman's statement something of a moot point, despite its clear alignments with the capitalist mode of production. However, neo-liberalism has certainly had a significant impact on CSR. Its transformation of organisations from "...perceived evil empires to 'partners' who could work with governments and non-governmental organisations to arrive at a more socially aware business environment" (Whitehouse, 2003:303) in turn demands some demonstrable actions and activities from such 'partners', given their new-found legitimisation. Indeed, it has been suggested that organisations now need a public "licence to operate" (Idowu & Louche, 2011) for their continued success, and this is where CSR admirably fills the gap. It can be used by organisations to demonstrate that they are a sustainable business that adds economic, environmental and social value, and are committed to behaving ethically while improving the quality of life for the workforce, their families and wider society (Baptiste, 2008).

The expectations of contemporary society and its demands for radical transparency and honesty in all corporate activities (Jones, 2012) also have an influence on organisational activity. Such expectations can be considered through more conceptual notions of CSR through which its development can be considered according to various emergent stages. Early Business In The Community (BITC) work proposed two distinct 'ages' of CSR: the 1990s termed 'show me',' and the 2000s 'prove to me' (Murray & Dainty, 2009), representing the change in social demands on organisations. More recently, Jones (2012) proposed that the 1990s were the "age of image", where CSR remained relatively superficial, the 2000s were the "age of advantage", where organisations brought about genuine change in order to enhance competitiveness, and most recently, the 2010s are designated the "age of damage", where those that do not meet expectations will suffer. Indeed, business ethics are becoming a more important aspect of organisational strategy as a whole, as bad publicity can now readily and rapidly affect sales and profits (Baptiste, 2008).

This has naturally led to a desire for organisations to closely control their CSR publicity and to create and manage its enactment and manifestations in practice to ensure and enhance brand value and reputation (BITC, 2011). This has itself been regarded as "...an illegitimate and dangerous transformation of social and environmental issues into commodities..." (Brés & Gond, 2014: 2), the commodification of worker HSW being only one such issue.

Contemporary CSR is arguably a product of the technological developments and trends in society (Pedersen, 2015). The rise of social media, and indeed the Internet as a whole, have had significant impact on how CSR 'works' in practice: the way we now communicate and the way in which corporate information and activities can be communicated means that organisations need to ensure a relevant and active profile and presence within this virtual space. Indeed, to return to BITC's ages of CSR, the 2010s could now be termed 'bombard me'. In positioning CSR as an essential organisational activity, it has now become something to be promoted, true propaganda in a "...consistent, enduring effort to create or shape events to influence the relations of the public to an enterprise" (Bernays, [1928] 2005: 52). As Herman and Chomsky (1988) argue, such PR propaganda essentially seeks to dictate society's expectations, to define the questions to be asked and the parameters to be judged, and as such is instrumental in what society expects CSR to look like. Indeed, Ennals (2011: 146) notes that in the US and the UK, "social responsibility" has been more concerned with optional corporate philanthropy where selected additional activities are placed in the public eye under the management of PR staff, than more discrete approaches to CSR in practice.

This brings with it several considerations. Organisational control over what is promoted as their CSR activity should be explicitly acknowledged, as CSR cannot always bring good news and research has shown "...revealing insights into what is, and is not, recorded in their annual reports" (csrnetwork, 2008: 236). The gap between CSR rhetoric and organisational practice has been termed 'greenwashing' (Pedersen, 2015), suggesting a heavy coat of PR gloss is often applied to bring the two back into some alignment. That organisations can also readily control their virtual images through manipulation of algorithms and other mechanisms to 'bury' bad news within the virtual world, while ensuring other stories 'bounce' to the top of search engines is also worthy of note (Merritt, 2016). The fact that CSR can be considered one of the key contributors to organisational PR 'presence' is perhaps more significant. Contemporary social media requires a certain amount of content to maintain volume and timely issue of tweets, posts or press-releases, and so organisations are encouraged to chase quick and easy 'wins', to undertake CSR activities and initiatives that readily generate frequent and cosmetic content to feed the insatiable PR machine.

This brief review of CSR has deliberately avoided any discussion of what constitutes CSR in practice. Rather, the context for CSR has been explored to reveal a marketplace where it has become an essential part of operations under the neo-liberal doctrine. Organisational benevolence has become a necessary consideration for organisational strategy – organisations *do* need to do good to do well – cemented by the role social media now plays in our world in shaping the demands and expectations of a much more rapidly informed society. As a result, organisations pro-actively promote their CSR through various channels to maintain brand integrity, yet an inevitable consequence of this is the need for an almost constant output of CSR activities that can be readily commodified and posted to the public as evidence of expected CSR, despite more critical concerns of superficiality, greenwash and the lack of effective legacy.

In practice, CSR is strongly influenced by this wider context, and this chapter will now firstly explore some specifics of CSR within the construction industry, before looking at one aspect in particular: the HSW of construction workers.

Corporate social responsibility in construction

It has been suggested that CSR is particularly important for construction industry organisations simply because their activities impact directly on society, the economy and the environment – the "three pillars of CSR" (Murray & Dainty, 2009; Chan & Cooper, 2011; Upstill-Goddard et al., 2012). It has become a mainstay of a construction business in the developed world: large construction organisations champion their values and their acts of corporate citizenship through their webpages and press releases (Sherratt, 2016a). Construction clients are keen to be associated with responsible construction firms and in the UK the Social Value Act 2012 now requires public authorities to consider economic, social and environmental well-being when awarding public sector contracts, embedding CSR in the guise of sustainability values within the construction procurement process.

There has been a shift in corporate governance thinking to include the notion of "giving something back" to the wider community (Chan & Cooper, 2011) in which the construction project is taking place which has resulted in many different construction CSR initiatives, for example, those involving local apprenticeships, free contributions to community construction projects and environmental activities. Perhaps unsurprisingly, given the construction industry's 'problematic image', significant efforts are being directed to generate, evidence and support organisational commitments to CSR in practice, as well as exploring how commitment to CSR can be harnessed to bring positive change to both the industry and society as a whole.

Much of the output of construction industry CSR reflects the relative superficiality needed for PR. Indeed, the employment of CSR consultants does not always seek out potential pro-active CSR activities but instead seeks to assess the likelihood of inspection, and enforcement should damage occur: "It is not so much a matter of complying with the law, but of avoiding being found out not to have complied" (Ennals, 2011: 149). This is where Friedman's "rules of the game" become relevant, and where the UK construction industry was recently revealed to have set some rather unpalatable rules of its own. The blacklisting scandal broke in 2009 when The Consulting Association (TCA) was found to be operating a blacklist of over 3,000 construction workers, containing information about their union activities, political views, and other details. This list was used by many large contractors operating in the UK construction industry to vet workers on their sites, and in many cases resulted in their exclusion and inability to find work for many years. A total of £75m has been paid in compensation to blacklisted workers, although without admission of liability in any court, with claims brought against Carillion, Balfour Beatty, Costain, Kier, Laing O'Rourke, Sir Robert McAlpine, Skanska UK and Vinci (BBC, 2016a). These are all major contractors operating in the UK, who also proudly champion their more palatable and ethical CSR activities on their websites.

A further point of relevance here is also the way the construction industry is structured, both nationally and globally. Construction is driven by supply chains, and these are considered necessary to manage fluctuations in workload and the need for organisational agility, yet supply chains also impact how CSR can work in practice. While good CSR practice requires that organisations should ensure that they hold their supply chains to the same levels of account (Panayiotou & Aravosis, 2011: 58), this has proved notoriously problematic across all industries. Indeed, major construction projects often have global supply chains and contractors working overseas may instead employ local rather than corporate CSR standards, something Smallwood and Lingard (2009) suggest is unethical. And again it is all too easy to find a construction industry example. Supply chain 'problems' have been clearly illustrated by the construction work for the FIFA 2022 World Cup in Qatar where worker exploitation has received a close examination from organisations such as Amnesty International (2016) who have called for all construction firms operating in the country to increase their efforts to ensure the protection of worker rights. UK construction firms working on the projects have been drawn into this debate, notably Carillion who were "deeply concerned and surprised" by BBC findings of poor working conditions on its sites in the region (BBC, 2014). Although the debate around whether workers are dying as a direct result of the World Cup continues, that construction fatalities in Qatar for Nepali and Indian workers *alone* totalled 1,239 deaths in the three years to the end of 2013 is a very stark statistic (BBC, 2016b) and evidence of poor worker HSW practices across Qatar.

It has been argued that with increasing societal awareness of CSR, the company workforce can become a driving force for organisational CSR strategy (Vogel, 2005): their embarrassment at working for a company involved in such exploitation can create a desire for change. Indeed, expectations are felt to be even greater in the younger generation, who are seen as a key influence in the development of worker human resource management policies (Christy, 2015). But supply chains frequently fracture and obscure this relationship; the exposure of practices in Qatar is perhaps the exception rather than the rule. In fact, although many large construction organisations champion worker welfare and HRM practices that align to CSR initiatives, it must be remembered that their direct employees are often *not* the people that build our buildings. Often cascading supply chains of gradually smaller contractors are involved, with sole traders or labour-only providers found working on large projects as subcontractors of subcontractors of subcontractors. This workforce remains invisible – their names are not found on the hoardings, and in reality, a large site might only have a very small percentage of people from that branded organisation working there: an office staff of perhaps 20 to a site of 500 workers. Therefore many CSR policies are only impactful to white-collar managers; as Ness (2009: 647) points out, there is often a "...denial of responsibility for the supply chain, ignoring the actual site operatives who do the work", in cases where HSW is commodified in practice.

While construction organisations also operate within the same global and economic context as set out previously; they also have some unique aspects to consider about CSR in practice. That CSR has become embedded in procurement routes to some extent secures its place in organisational strategy, and indeed through legislation such as that set down by the UK's largest construction client, the UK government. However, for construction this makes things all the more embarrassing, given the scale of some recent unethical practices by UK contractors, both at home and abroad. The role of supply chains is also fundamental here, both in the way CSR commitments can travel up and down them robustly, but also with regard to who actually comprise 'the workforce' who may benefit from CSR in practice, particularly when it attempts to bring positive change to such important aspects as worker HSW.

Is worker HSW a corporate social responsibility?

HSW has certainly become entrenched in organisational CSR (Sherratt, 2016a), often seen as sitting on the "social pillar" of the "triple bottom line" (Panayiotou & Aravosis, 2011). Indeed, Smallwood and Lingard (2009) have argued that HSW is an *essential* component of CSR, in the way that CSR can ensure corporate awareness of and commitment to HSW in practice, in turn leading to enhanced overall performance and reduced cost (Smallwood & Lingard, 2009: 282), and aligning the "business case" for HSW with that of CSR. They also note that HSW needs to be instilled as an organisational "value" rather than a "priority" to ensure its resistance to change and inclusion alongside other key performance parameters (Smallwood & Lingard, 2009). However, the subjugation of HSW to contemporary CSR practice has arguably mutated such honourable intentions.

That CSR is also concerned with PR, brand image, and corporate presence has significant repercussions for this relationship, which further removes worker HSW from production in the workplace. Indeed, Friedman (1962: 133) queried how businessmen could know what CSR should be or even look like, beyond making profits for their shareholders. The same question could well be asked of

PR professionals as well. When HSW becomes a CSR 'activity', the composition of the decision makers also changes, from those at the 'sharp-end', where practical management is only all too aware of the catastrophic repercussions if something goes wrong, to those far removed from the 'muck and bullets' of the site environment. And although society now has certain demands of organisations, albeit that these are to a large extent shaped by CSR PR itself, it is very unlikely that they also have the skills to unpick practical HSW management activities, particularly when many construction professionals still struggle to produce a worthwhile and relevant method statement to set out and plan work safely on sites (Borys, 2012).

But they do understand metrics. The nice big round target zero of many contemporary construction safety programmes proves a popular choice for its visual, absolutist and readily comprehensible appeal, despite many problems in its practical implementation such as worker disenchantment and statistical vulnerabilities (Sherratt, 2014). Targets can also easily be assigned to health and well-being activities, for example, the percentage of workers on sites who have been checked for certain health conditions, or the percentage of workers on sites who have committed to stopping smoking. Indeed many CSR 'wellness measures' are actually about prevention of disease rather than a promotion of health, but they still inevitably label individuals, stigmatising lifestyle choices and creating personal guilt (Conrad, 2005). It is also often assumed that people are happy to participate in health surveillance schemes, but this is not always the case. Many companies and individuals opted out of the occupational health programme on the UK Olympic 2012 project (Tyers & Hicks, 2012), reflecting Illich's (1976: 99) suggestions that screening and testing simply "…transforms people who feel healthy into patients anxious for their verdict".

The vast majority of HSW simply *cannot* be measured (Townsend, 2014). Safety is itself a non-event (Hollnagel, 2014 after Weick) while health and well-being often defy measurement owing to their emergence through a "slow-burn", and indeed the fact that they can be much more problematic to manage in practice (Sherratt, 2016b). But this is very inconvenient for CSR, which is itself drawn to metrics for the measurement required for continuous improvement and verifiable corporate citizenship. Indeed, De Colle et al. (2014) have suggested that the use of CSR standards can be highly problematic, and go so far as to consider it a "paradox of CSR" that such standards may create unintended consequences precisely *because* of their need for measurement. As it is impossible to measure intangible aspects directly, proxies are used, but such proxies "…can lead to a progressive loss of attention to the important, but unmeasurable aspects" (De Colle et al., 2014: 184) and so a shift to the measurable but relatively superficial emerges: the daily provision of 'healthy' bowls of fruit in the site canteen superseding the much more complex task of managing the occupational health hazard of dust control across the project. CSR reporting requires simple, straightforward and ideally photogenic content, not lengthy articulations of complicated and unattractive management practice. The inclusion of HSW under the CSR umbrella inevitably panders to the former; everyone knows fruit is healthy and colourful and able to provide much more visually stimulating content than the provision of face-fit dust masks – for one thing, the latter inconveniently obscure the smiles of the workforce modelling them.

And this links to a further challenge which relates to one of the fundamental tenets of CSR itself: the desire to go *beyond* minimum legal requirements in corporate operations (Whitehouse, 2003; Murray & Dainty, 2009; Panayiotou & Aravosis, 2011). In many countries worker HSW is enshrined in legislation. In the UK, for example, the Health and Safety at Work, etc. Act 1974 forms the core foundation of all HSW legislation and is supported by various regulations. These set out the requirements for, amongst many others, the need to conduct suitable and sufficient risk assessments, to provide personal protective equipment (PPE), and to programme medical screening for at-risk workers. For those working outside of construction, health and safety are often considered to be "… legally constructed and largely focused on risk assessment and the prevention of work-related injury or illness" (Christy, 2015: 78). Instead, they turn to well-being as the aspect that goes beyond such minimum legal thresholds, and so this has become most prominent within CSR activities. For the construction industry, where the presence of legislation certainly does not equal compliance (as many accident investigations sadly demonstrate), the PR machine often neglects to mention *any* need for regulatory compliance within their presentation of CSR. They prefer instead to give the impression that companies are voluntarily going 'above and beyond' in their commitment to their workforce HSW, for example, in the provision of 'site health checks', when in fact, all they are doing is simply meeting minimum legal requirements.

It must be acknowledged that these criticisms do not seek to negate the efforts of those managing construction work in practice – the site supervisors and managers who are tasked with ensuring worker HSW every day on sites: such mundane practice (in the eyes of PR) is unlikely to ever become fodder for the CSR machine. But what is of concern is the shift in rhetoric and indeed corporate practice to worker HSW following its inclusion within the CSR remit. Indeed, that this phenomenon exists at all suggests misdirection of corporate efforts that could perhaps be better spent on improvements in construction technology or practice – activities that could go some way in preventing the many negative effects of construction work on the construction worker occurring in the first place. Should construction worker HSW be a corporate social responsibility in the form that CSR currently takes in practice? Perhaps not. The influence of CSR on worker HSW has to some extent inevitably been informed by the requirements of PR, and consequently reduced to the readily commodifiable – to content that can be easily understood and appreciated by the widest possible audience, and so comes to shape their expectations. Metrics have become prominent, despite their problematic associations with HSW, and indeed the potential for their paradoxical influences on CSR in general. Workers also remain vulnerable to various negative associations of this commodification of their HSW if they are even impacted by CSR activities, given their place in the supply chain. It can even be argued that given the current capitalist mode of production, and the inevitable commodification of worker HSE within this context, CSR is inevitably exercised in favour of organisations to the detriment of their workers. To explore these challenges in more depth and provide empirical support for the arguments made here, a case study of UK construction industry health is now presented.

Case study: UK construction industry health

This case study is based on a UK government initiative – the Public Health Responsibility Deal launched in the spring of 2011. Commercial organisations sign up to the deal and its associated pledges to demonstrate their commitment to supporting the public health of the nation in an explicit act of CSR. This neo-liberal project sought to shift public health into the occupational domain, seeking industry partners to help the UK government develop a "...more collaborative approach to tackling the challenges caused by our lifestyle choices" (Department of Health [DoH], 2013a). The Health at Work Pledge, which itself originally contained four 'sub-pledges' (chronic conditions guide, occupational health standards, health and well-being reporting and the provision of healthier staff restaurants) was eventually supported by a specific Pledge H10 for Construction and Civil Engineering Industries, launched in the autumn of 2013 (DoH, 2013b).

An explicitly critical approach is taken to the primary data that forms this case study, predominantly employing critical discourse analysis (Fairclough & Wodak, 1997; Gergen, 2009) to explore the pledge through the text of its accompanying press release and the corporate websites of the 'top-ten' UK construction organisations that are able to reflect its impact on industry practice (later work carried out in 2016). Consequently, no claim is made to generalisation beyond this case study, but it is hoped that readers can find both fit and resonance with their experiences and practices of HSW, and the growing influence of CSR throughout the global construction industry.

The Construction and Civil Engineering Industries H10 pledge was launched with a press release titled: *Britain's beefy builders say bye bye to baring bottoms* (DoH, 2013a). The title alone clearly demonstrates what the PR machine thinks of the construction workforce, and this attitude permeates throughout the press release document. Construction workers are simply positioned as a homogeneous human resource, their individual identity and indeed ownership of their health are only championed in a few soundbites from UK TV Celebrity Builder Tommy Walsh who argues that the average construction workers are "...just as likely to go down to the gym as they are to go to the pub after work". This voice remains unique in a discourse which essentially negates the worker, and employs the stereotypical construction of the 'bottom-baring, overweight builder' to provide justification for the involvement of construction organisations in their workers lives: to 'support' and 'care' for the workers and their health.

The links to Victorian ideals of philanthropy and indeed some of the roots of CSR are clearly evident here, but it must be acknowledged that in asking organisations to help their "workforce lead healthier lives" a significant boundary has been crossed: organisations are given legitimate control of worker health beyond the site gates to the full extent of their working lives. This is inherently problematic in an industry with a poor occupational health record, and to shift perspective from the workers at work to the workers in their whole life creates a change in both the ownership and management contexts of such health problems. For example, the incorporation of smoking into a worker's health profile could potentially change the liability for any future lung disorders, whether the company paid for correctly face-fitted dust masks or not.

The invocation of benevolent paternalism from construction organisations also raises concerns. In suggesting that workers need help in their lifestyle and health choices, and indeed that they make poor decisions around what Thaler and Sunstein term "sinful goods" (2008: 80) such as "...smoking, alcohol and jumbo chocolate doughnuts" interferes with worker choice under the guise of pastoral power (Foucault, 1982, 2002). Suggesting that there are clearly defined 'higher' (ideal and rational) and 'lower' (impulsive and uncontrolled) selves, Berlin (1958: 18) argued that this kind of language leads to the rationalisation and justification of "...coercing others for their sake..." towards goals that they "...would, if they were more enlightened, themselves pursue". But for Friedman's businessmen or the PR professionals now in charge of HSW through CSR, to make such decisions for their workforce means corporations are now fundamentally challenging worker autonomy and personal freedoms. Berlin describes equating what people would choose if they were something they are not, with what they sought and chose as a "...monstrous impersonation... at the heart of all political theories of self-rationalisation" (1958: 18). This lack of worker rationalisation is something found throughout the Pledge: the 'overweight builder' unable to realise his true interests because "...many construction workers do not have ready access to the kind of general support to promote better health that is available to other working people". This example of a highly simplistic and indeed patronising rationalisation lies at the heart of the wider discourse that legitimises workers' health as the responsibility of others, specifically their employers, further reinforcing its position under the umbrella of CSR.

However, praise for organisations "making the health of their staff a priority on their sites" rings rather hollow when considered within the actual UK legislative context, which states that this is *not* actually a voluntary situation – they should be doing precisely that anyway: they are claiming to go beyond when in reality only minimum requirements are being met. Such obfuscation is welcomed and indeed perpetuated by the organisations themselves as they reflect the language and contextualisation of this press release in their organisational approaches. For example, one organisation stated on its website that "...all our businesses will conduct health checks and health risk assessments to ensure there is no long-term harm to health from working in our business", again demonstrative of the practice of positioning of risk assessments and health surveillance as pro-active efforts and demonstrable CSR, rather than the minimum legal standards they actually are (Health and Safety Executive, 2015b).

Within the websites, the impact of the pledge and its overall approach to construction health is clear. While occupational health plays second fiddle to safety regarding rhetoric and prioritisation in content; public health has blossomed as a fundamental organisational 'concern', frequently linked with well-being, as something 'supported' by the organisation but without any corresponding level of detail or operational specificity. Explicit references to the Public Health Responsibility Deal can be found, with organisations demonstrating their commitment to the pledge through the presentation of detailed health and well-being programmes seeking to educate and encourage lifestyle changes, or simply through core 'value' statements that "we support health and well-being". Such statements clearly reflect PR in practice, the ability to set the expectations of society around something both vague and meaningless (Herman & Chomsky, 1988), helping to limit the need for more significant corporate action. The organisations have readily adopted their assigned role of a benevolent shepherd to their workforce flock; however, the complexities of supply chains within the construction industry are also evident, and who this workforce is remained much more obscure. While reference is made to "everyone who works with us", in only one case was the supply chain specifically highlighted; more often 'all our people' provides a comfortable lack of specificity in the practical implementation of the health and well-being programmes. Indeed, whilst one organisation's aim is for "…75 percent of our employees [to be] using the programme by 2020", this will in reality be far less than the number of workers that will actually contribute to their construction outputs on sites.

Organisations are also happy to highlight public health 'problem issues', which include diabetes, high blood pressure, and stress, while those seeking promotion include active lifestyles, smoking cessation and healthy eating with offers of on-site wellness screening and health-screening clinics with the overall aim of 'building a healthier workforce'. Although many of these issues relate to individual choice and lifestyle, areas where notions of autonomy and personal freedoms again become highly significant, they are also areas that enable the organisation to distance themselves from their manifestation within the workforce. These are individual issues, not organisational, problems of public, not occupational health, and as such the organisation can confidently take the benevolent position, helping the workforce make the 'right decisions' about their health and lifestyles, readily providing demonstrable yet relatively inert CSR in practice. And if organisations are 'committed to promoting healthier lifestyles and helping our people to manage their health', surely they are also doing all they can for occupational health in their workplaces already? But that can be readily challenged when the overall occupational health record of the UK construction industry is examined in any detail.

One of the biggest impacts of the incorporation of HSW under the CSR umbrella is related to the actual location of much of the health data beyond HSW webpages. Only three of the ten organisations shared a formal HSW policy via their websites; far more chose instead to champion their HSW activities within press releases or news articles; health (and safety and well-being) neatly packaged and photographed to create PR content. For example, one organisation had "…agreed to purchase fresh fruit from a local stall holder to provide fruit for our operatives…" as part of their healthy eating drive; another had run "…local awareness campaigns around areas such as mental health and well-being, healthy eating and drug and alcohol abuse…" which had resulted in a Better Health at Work Award. Indeed, the number of awards that can be won in this area of organisational management is quite impressive, and although safety still dominates, health remains much more prominent in its attractive CSR-friendly public form than any associations with occupational health in practice. This lends clear support to the argument that such ventures are indeed only attempts to continue to feed the ever-hungry PR machine.

More seriously, however, it can also be suggested that this 'packaging of health' as revealed by the data has influenced the emphasis on public health over occupational, and consequently shaped the wider discourse of health within construction. It has set expectations of what worker HSW *should* look like in the construction industry, and enables this superficiality to continue in perpetuity as society's demands are inevitably met by future initiatives, activities, and output. The commodification of worker health into PR packages and awards, and by association, the organisations' ability to position themselves as benevolent champions of the workforce (although which one remains somewhat unclear), also enable industry clients to align themselves with a very attractive version of the UK construction industry. The development of this attractiveness, something now expected by our 'bombarded' society, is arguably a misdirection of investment, efforts, and practices to feed the superficial and photogenic at the expense of the neglect of occupational health within the UK construction industry.

Summary

In the introduction of this chapter, it was asked whether CSR was the ideal home for something as fundamentally ethical as worker HSW. Surely worker HSW *is* a corporate social responsibility, something at the very heart of 'doing good' to whatever end. The answer that has emerged is both 'Yes' and 'No' – yes, worker HSW is a corporate responsibility, and as people are involved, why not make it social too, but also no, at least not in the way CSR currently works within our contemporary society.

The need for organisations to maintain their online presence, to seek out positive content that can be readily understood by society, to bolster their CSR credentials, which in turn enhance an organisational image and therefore work-winning potential, all indicate that CSR is not the ideal home for worker HSW. As the case study demonstrates, the temptation to grab the photogenic 'low-hanging fruit' of worker public health initiatives to feed CSR content is too much for many large construction companies to pass by. The compulsion to dress up straightforward legislative compliance as corporate citizenship to further shore up their commitments to both their workforce and wider society also seems somehow impossible to resist. By positioning worker HSW within the remit of

CSR it essentially becomes a commodity, and in contemporary society that means something tweetable or post-able, which has, in turn, enabled the shaping and control of expectations of worker HSW by society. We have been conditioned by the growing influence of PR and CSR on worker HSW to expect work wellness programmes and healthy on-site initiatives, despite the fact that they are essentially a distraction from the much more serious issue of managing worker occupational HSW in practice. Indeed, CSR enables construction organisations to play the role of a benevolent shepherd to their workforce flock, superficially supporting them in weight loss or smoking cessation, all the while demanding their blood, sweat, and tears as they labour through poor occupational health conditions out on sites, their personal freedoms challenged and constrained.

The question must now be asked: if this is wrong, what would be right? And as noted above, yes, worker HSW *is* a corporate responsibility, but it is one that should never be influenced by what CSR needs for its manifestation and perpetuation. Worker HSW must be removed from the ownership of the businessmen and the PR professionals and placed firmly back on the sites – where the adverse effects of poor HSW management are felt, and the suffering occurs. The inevitable commodification of worker HSW within the construction site context should be explicitly acknowledged. Easily measurable metrics should be avoided and should not replace an understanding of the real complexity of HSW management in practice, while legal requirements should be met in full, rather than dressed up in CSR finery for future 'sale'. Indeed, the investment of CSR publicity budgets into work contracts would perhaps be a good starting point. For example, ensuring that suitable mechanical means for lifting is readily provided on all sites is preferable to superficial encouragement for workers to 'healthily' develop the muscles needed for repeated manual handling in practice. The *real* interests of construction workers are likely to be a balance of good occupational and public HSW management, but what this looks like should not be dictated by government, much less commercial organisations with vested interests in worker output and the ability to shape the discourse to any significant extent.

While there may well be potential for wider industry reform through CSR, and there may be a margin of common interest that can be exploited before the hidden conflicts become manifest, it is arguable that the construction industry is simply not yet there regarding its fundamental work practices and occupational activities. It has become distracted by the ready fit of HSW under the CSR umbrella, despite the problems this also brings. Indeed, why would companies wish to change the way they work on sites using long supply chains, readily able to obfuscate the realities of actual construction worker HSW when they are so readily able to shape expectations and then claim corporate citizenship through much more superficial means?

The industry needs to embrace the ideas of valuing people in construction in its operations, looking to address the fundamental issues of occupational HSW in practice, before looking beyond this to try to change the world – if indeed that can ever be possible should the commodification of construction worker HSW continue unchallenged.

Acknowledgements

The author would like to thank Andrew Hale for the interesting discussions and communications that have contributed to this work; also Barry Rawlinson, Esther Norton, and the anonymous reviewer of an earlier draft of this chapter, for their time and input. The case study presented here draws on research first published in *Construction Management and Economics* and by the Association of Researchers in *Construction Management* (Sherratt, 2015; Sherratt, 2016a).

References

Amnesty International (AI). (2016), *The ugly side of the beautiful game – Exploitation of migrant workers on a Qatar 2022 World Cup site*, Amnesty International [Online]. Available at: http://bit.ly/2b3FWuk Date of access: 3 August 2016.

AI *see* Amnesty International

Baptiste, N. R. (2008), The symbiotic relationship between HRM practices and employee well-being: A corporate social responsibility perspective. In: D. Crowther & N. Capaldi (eds), *The Ashgate research companion to corporate social responsibility*, Ashgate Publishing, Farnham, pp. 151–180.

BBC *see* British Broadcasting Corporation

Berlin, I. (1958), *Two concepts of liberty*, Oxford University Press, Oxford.

Bernays, E. (1928, 2005), *Propaganda*, Ig Publishing, New York.

BITC *see* Business in the Community

Borys, D. (2012), The role of safe work method statements in the Australian construction industry, *Safety Science*, Vol. 50(2), pp. 210–220.

Brès, L. & Gond, J.-P. (2014), The visible hand of consultants in the construction of the markets for virtue: Translating issues, negotiating boundaries and enacting responsive regulations, *Human Relations*, Vol. 67(11), pp. 1347–1382.

British Broadcasting Corporation (BBC). (2014), *Qatar 2022: Construction firms accused amid building boom* [Online]. Available at: http://bbc.in/2aJc4dX Date of access: 3 August 2016.

British Broadcasting Corporation (BBC). (2016a), *Construction workers win payouts for 'blacklisting'* [Online]. Available at: http://www.bbc.co.uk/news/business-36242312 Date of access: 3 August 2016.

British Broadcasting Corporation (BBC). (2016b), *Have 1,200 World Cup workers really died in Qatar?* [Online]. Available at: http://bbc.in/2aSiJaY Date of access: 3 August 2016.

Business in the Community (BITC). (2011), *The business case for being a responsible business*, Cranfield University School of Management, Cranfield.

Chan, P. & Cooper, R. (2011), *Constructing futures: Industry leaders and futures thinking in construction*, Wiley Blackwell, Chichester.

Christy, G. (2015), CSR and human resource management. In: E. R. G. Pedersen (ed.), *Corporate social responsibility*, Sage, London, pp. 72–102.

Conrad, P. (2005), Wellness in the work place: Potentials and pitfalls of work-site health promotion. In: P. Conrad (ed.), *The sociology of health and illness: Critical perspectives* (7th edn), Worth Publishers, New York.

csrnetwork. (2008), CSR reporting – examining the unpalatable issues. In: J. Burchell (ed.), *The corporate social responsibility reader*, Routledge, Abingdon, pp. 236–238.

De Colle, S., Henriques, A. & Sarasvathy, S. (2014), The paradox of corporate social responsibility standards, *Journal of Business Ethics*, Vol. 125(2), pp. 177–191.

Department of Health (DOH). (2013a), *Britain's beefy builders say bye bye to baring bottoms* [Online]. Available at: http://bit.ly/PAoK0R Date of access: 9 August 2016.

Department of Health (DOH). (2013b), *Health at work network* [Online]. Available at: http://bit.ly/1kXxYOw Date of access: 9 August 2016.

DOH *see* Department of Health

Ennals, R. (2011), Labour issues and corporate social responsibility. In: S. O. Idowu & C. Louche (eds), *Theory and practice of corporate social responsibility*, Springer, London, pp. 143–158.

Fairclough, N. & Wodak, R. (1997), Critical discourse analysis. In: T. A. van Dijk (ed.), *Discourse as social interaction*, Sage, London.

Foucault, M. (1982, 2002), The subject and power. In: J. D. Faubion (ed.), *Michel Foucault: Power - essential works of Foucault 1954–1984*, Vol 3, Penguin, London, pp. 326–348.

Friedman, M. (1962), *Capitalism and freedom*, University of Chicago Press, London.

Friedman, M. (1970), The social responsibility of business is to increase its profits, *The New York Times Magazine*, September 13.

Frynas, J. G. (2009), *Beyond corporate social responsibility*, Cambridge University Press, Cambridge.

Gergen, K. J. (2009), *An invitation to social construction* (2nd edn), Sage, London.

Health and Safety Executive (HSE). (2015a), *Health and safety in construction sector in Great Britain* 2014/15 [Online]. Available at: http://bit.ly/Sgu84d Date of access: 9 August 2016.

Health and Safety Executive (HSE). (2015b), *Is health surveillance required in my workplace?* [Online]. Available at: http://bit.ly/1QDzp5O Date of access: 9 August 2016.

Henderson, D. (2001), *Misguided virtue – false notions of corporate social responsibility*, Institute for Economic Affairs, London.

Herman, E. S. & Chomsky, N. (1988), *Manufacturing consent – the political economy of the mass media*, Vintage, London.

Hollnagel, E. (2014), *Safety I and Safety II – The past and future of safety management*, Ashgate Publishing, Farnham.

HSE *see* Health and Safety Executive.

Idowu, S. O. & Louche, C. (2011), *Theory and practice of corporate social responsibility*, Springer, London.

Illich, I. (1976), *Limits to medicine – medical nemesis: The expropriation of health*, Penguin Books, London

Jones, D. (2012), *Who cares wins – why good business is better business*, Pearson Education, Harlow.

Marx, K. (1976, 1990), *Capital* Vol 1. Translated by Livingstone, R., Penguin Books, London.

Merritt, J. (2016), *Reputation management 101: How to remove negative search results* [Online]. Available at: http://bit.ly/2aNuZMu Date of access: 11 August 2016.

Murray, M. & Dainty, A. R. J. (2009), Corporate social responsibility: Challenging the construction industry. In: M. Murray & A. R. J. Dainty (eds), *Corporate social responsibility in the construction industry*, Taylor and Francis, Abingdon, pp. 3–23.

Ness, K. (2009), Not just about bricks: The invisible building worker. In: A. R. J. Dainty (ed.), *Proceedings of the 25th Annual ARCOM Conference*, 7–9 September 2009, Nottingham, UK. Association of Researchers in Construction Management, Vol. 1, pp. 645–54.

Panayiotou, N. & Aravosis, K.G. (2011), Supply chain management. In: S. O. Idowu & C. Louche (eds), *Theory and practice of corporate social responsibility*, Springer, London, pp. 55–71.

Pedersen, E. R. G. (2015), The anatomy of CSR. In: E. R. G. Pedersen (ed.), *Corporate social responsibility*, Sage, London, pp. 3–36.

Rawlinson, F. & Farrell, P. (2010), UK construction industry site health and safety management: An examination of promotional web material as an indicator of current direction, *Construction Innovation, Information, Process Management*, Vol. 10(4), pp. 435–446.

Sherratt, F. (2014), Exploring 'Zero Target' safety programmes in the UK construction industry, *Construction Management and Economics*, Vol. 32(7–8), pp. 737–748.

Sherratt, F. (2015), Legitimising public health control on sites? A critical discourse analysis of the Responsibility Deal Construction Pledge, *Construction Management and Economics*, Vol. 33(5–6), pp. 444–452.

Sherratt, F. (2016a), Shiny happy people? UK construction industry health: Priorities, practice and public relations. In: P. Chan (ed.), *Proceedings of the 31st Annual ARCOM Conference*, 5–7 September 2016, Manchester, UK, Association of Researchers in Construction Management.

Sherratt, F. (2016b), *Unpacking construction site safety*, Wiley-Blackwell, Chichester.

Smallwood, J. & Lingard, H. (2009), Occupational health and safety (OH&S) and corporate social responsibility. In: M. Murray & A. R. J. Dainty (eds), *Corporate social responsibility in the construction industry*, Taylor and Francis, Abingdon, pp. 261–286.

Thaler, R. H. & Sunstein, C. R. (2008), *Nudge – improving decisions about health, wealth and happiness*, Penguin, London.

Townsend, A. S. (2014), *Safety can't be measured: An evidence-based approach to improving risk reduction*, Gower Publishing, Farnham.

Tyers, C. & Hicks, B. (2012), Occupational health provision on the Olympic Park and athletes' village. *Health and Safety Executive Research Report 921*, HSE Books, Norwich.

Upstill-Goddard, J., Glass, J., Dainty, A. R. J. & Nicholson, I. (2012), Integrating responsible sourcing in the construction supply chain. In: S. D. Smith (ed.). *Proceedings of the 28th Annual ARCOM Conference*, 3–5 September 2012, Edinburgh, UK, Association of Researchers in Construction Management, pp. 1311–19.

Vogel, D. (2005), *The market for virtue – the potential and limits of corporate social responsibility*, Brookings Institution Press, Washington DC.

Whitehouse, L. (2003), Corporate social responsibility, corporate citizenship and the global compact: A new approach to regulating corporate social power? *Global Social Policy*, Vol. 3(3), pp. 299–318.

13 Making zero harm work for the construction industry

Fidelis Emuze and Fred Sherratt

The best way to predict the future is to invent it

(Immanuel Kant)

Background

Improving the health, safety, and general well-being of people in complex socio-technical systems, while preventing diseases, fatalities, and injuries, is a difficult task – and certainly not a *fait accompli* for the construction industry. Efforts are continually being made to seek improvements and help preserve the life of people in construction, and both researchers and practitioners have developed and promoted various safety management systems (SMSs), incorporating a multitude of different concepts, methods, tools and philosophies to eliminate workplace hazards and risks, with the intention of banishing harm. SMSs are documented in reports, policies, and safety manuals, and on the Internet, where contractors indicate their commitment to higher standards of health, safety, and well-being of their workers and the general public (Rawlinson and Farrell, 2010). However, the simplest search of the Internet will also attest to the fact that harm has not yet been removed from construction work.

The International Labour Organization (ILO) has long decried the harm done to people by the construction industry. For example, the ILO reported in 2008 that the construction fatality rate in some industrialised countries ranged from 3.3 to 10.6 deaths per 100,000 workers (CPWR, 2013). Eight years on from this report, the situation is not much different. Indeed, fatality rates have even escalated in some developing countries (Haslam et al., 2005; Kheni et al., 2008), as the volume of construction work has itself increased. Behind this statistical data are the lived experiences of people, each one in some way harmed by their work in the construction industry. While different people will naturally ascribe different meanings to fatal and non-fatal injuries encountered in the construction industry, harm itself generates some significant associations, as shown in Figure 13.1.

Clearly, harm is to be avoided wherever possible, and this should certainly be achievable within structured and managed work environments. This has, perhaps inevitably, led to the concept of zero harm within health and safety management.

Figure 13.1 Potential construction industry harm

Zero harm can be found in many different industries and wider societal initiatives, such as the Vision Zero safety programme for Scandinavian road safety (Swuste, 2012). However, there has also been the criticism of zero harm in practice, grounded in various theoretical concerns and the potential for unintended negative consequences of its adoption in practice (cf. Long, 2014; Sherratt, 2014). While not overlooking the merits within such criticisms, this chapter instead propounds the need to stop asking "Is the concept of zero harm an achievable goal in construction?", and it instead puts forward the suggestion that we should be asking "How can we make zero harm work for the construction industry?"

This chapter seeks to encourage acceptance of the *vision* of zero harm, a significant innovation in safety thinking (Zwetsloot et al., 2017), and to make an informed plea for dedicated research efforts geared towards its benefits and limitations within the construction industry context. By moving towards a zero harm vision that works for the construction industry, the industry can then start to make real differences in the health, safety, and well-being of the people in construction, those most vulnerable to harm.

The nature of the construction industry

The ever-increasing complexity of industrial systems often requires an approach to health, safety, and well-being that goes beyond common, rational explorations of technical systems, patterns, and procedures, to account for the active nature of processes and actions that influence the realisation of outcomes. Such growing complexity necessitates looking beyond the resolution of "tame problems", and instead demands to craft the right approach to solving "wicked problems" that always defy formulas (Hardin, 2009). As shown in Figure 13.2, tame problems are either known (scientific knowledge) or knowable (scientific approach) (Snowden, 2000), while wicked problems are complex (social systems) and chaotic.

Few would argue with the suggestion that many of the problems found in modern construction planning and execution are wicked in nature. About health and safety management, known issues have causes and consequences that are understood and can be anticipated, to the extent that decision-making only involves identifying risk, understanding the context, and applying best practices. While still slightly complicated, the causes and effects of knowable problems can be determined if enough data exists, and this can then be used to develop good practice as required. In contrast, complex and chaotic problems are not so easily tackled. In a complex context, causes and effects are always determined after the event. Situation analysis, exploration of alternatives, problem formalising, and flexibilities can help to arrive at the appropriate decisions, but such a system lends itself to post-incident studies (case studies, evaluations, and similar research). More significantly, the chaotic context has no constraints, to the extent that causes and effects cannot be identified, because making sense of situations can only come about from test runs of ideas and actions and subsequent observation of results. Again, the chaotic context resonates closely with some of the issues that are still to be resolved in construction management research, particularly in health and safety.

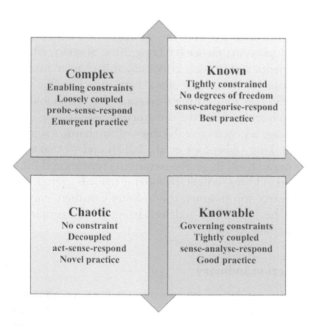

Figure 13.2 Illustrated domains of the Cynefin model (Adapted from Snowden, 2002: 108)

The predominance of the complex and chaotic in construction health and safety situations implies that more rigid, "straitjacket" solutions are failing the industry. Indeed, a shift away from such approaches is advocated, because:

> there is some evidence that making subsystems safer could make the overall system less safe because of the propensity of humans to take less care personally when a system takes more care.
>
> (French et al., 2011: 761)

In fact, working conditions, coupled with influencing factors (both external and internal) often increase the degree of complexity on safety encountered in industrial systems (Rosness et al., 2012). This is consistent with Snowden's (2000) contention that human reliability analysis models are akin to known and knowable contexts, but such models are unable to tackle complex and chaotic contexts. Perhaps the continued use of known and knowable safety models and approaches may even have contributed to the continued incidence of accidents, injuries, and fatalities, underpinned by poor work practices in the construction industry. Indeed, it is arguable that such sociotechnical contexts always produce complex and chaotic situations that are challenging, and so, multiple incidents and accident causations are also a major hurdle to be considered in such environments.

The BP incident of 2010 serves as a good example, particularly as the petrochemical industry is one that lays claim to superior SMS programmes, due to both the nature of the work and potential consequences of its operations. Complexity was identified in the BP accident report, as follows:

> The team did not identify any single action or inaction that caused this accident. Rather, a complex and interlinked series of mechanical failures, human judgments, engineering design, operational implementation and team interfaces came together to allow the initiation and escalation of the accident.
>
> (BP, 2010: 31)

Thus, multiple causal factors contribute to accidents, and the BP incident occurred in an industry in which large organisations are committed to goals in the form of "zero harm", "goal zero", and a "zero accident vision" (ZAV), etc. For example, the vision of Shell, another petrochemical organisation, is "no harm and no leaks", as part of their Goal Zero programme, which has been implemented across all its activities and businesses, and which is now a decade old.

A further example from Hafey (2015: 67) shows that:

> The incident that was investigated involved a forklift that slid on a wet floor in the shipping area of the plant. The initial focus of the investigators was the forklift driver and the speed at which he was driving. It was implied that he should have slowed down because he knew the floor was wet. By asking why five times I tried to steer them away from the person and toward the real cause.

Why did the forklift slide? Why was the floor wet? Where did the water come from? […] Are there different tires that would be more suited to wet conditions?

Further analysis of this incident moved the focus away from the "who" question, to the "what" and "why" questions, which, in turn, shed light on how the hazard should have been addressed in practice (Young, 2014).

These incidents show that simply setting goals is not enough in sociotechnical environments where causation models for hazard-related incidents emphasise the influence of organisational culture and the gaps in controls when policies, standards, procedures, and the accountability system fail (Manuele, 2014). Put simply, addressing problems with known and knowable models is not suitable in a complex and chaotic context. Instead, a more systemic and resilient approach is needed. Such an approach should have the ability to respond to usual and unusual threats in a robust and flexible manner, to observe and control the situation, including its performance, to anticipate risks and opportunities, and then to promote continuous learning from lived experience (Hollnagel, 2011; Hollnagel and Woods, 2006; Steen and Aven, 2011). It is possible that zero harm can contribute to such an approach and help create this step-change in health and safety management in practice.

Criticisms of zero harm

There are both proponents and opponents within the safety research literature regarding zero targets and the concept of zero harm. Criticisms range from the philosophical (Sherratt, 2014) to the ideological (Long, 2014; Dekker et al., 2015) to the more practical (Zwetsloot et al., 2017), and there is ongoing debate around zero harm within academic and industry-focused literature, and across industrial contexts. The key threads of the discussions that constitute the essential criticism of zero harm can be identified as follows:

- *The meaning of zero*: The potential for different meanings to be ascribed to the vision of zero harm by managers, operatives, and other practitioners is deemed to be a significant problem. The notion that all accidents, injuries, and harm in the workplace can be prevented is considered too simplistic when considered from within highly complex and chaotic sociotechnical systems. This sentiment is often found within the industry, and advocating the idea that "an accident is not a project requirement" sounds too much like impossible utopian thinking to conservative practitioners and those that work at the "sharp end" of industry practices (Long, 2014; Sherratt, 2014).
- *A numbers game*: The so-called "numbers game" is perhaps the most widely held criticism of zero harm, and it relates to reporting and accountability. The concern is that when zero is used as a target in safety controls, it can easily lead to under- and non-reporting of incidents, to achieve or maintain such targets. Zero harm programmes have the potential to preoccupy managers and supervisors with counting and shifting data because of personal and company gratifications, which can become a distraction from managing

safety in practice. This kind of situation creates a new range of language, spin and reporting dynamics to help concerned actors lay claim to target achievement, which can be tied to bonuses and other rewards. Excessive data capture and calculation (measurement), and high bureaucratic systems supported with regulations can also keep the distracting numbers game alive (Dekker, 2014; Hale et al., 2013).

- *Safety propaganda*: There is a common perception that indiscriminate use of the term "zero harm" has reduced the term to just a slogan, namely "safety propaganda", which simply fades away over time (Sherratt, 2016: 129). Such perceptions are often prevalent among frontline workers, who are perhaps rightly sceptical about safety initiatives, from their outset, and who are looking for evidence of tangible change and beneficial implications for workplace practice. When no positive change is evident in their workplace practices, initiatives such as zero harm are readily reduced to "just another safety slogan". This reaction of workers could be attributed to the widely held view that good safety initiatives never fade away, but perpetuate and develop over time, bringing improvements to workers' lived experiences.

- *Language effects*: Closely linked to ideas of safety propaganda, the way zero harm encourages people to talk about health and safety, and the discourse it develops, can also have an influence. There is the potential for disconnect between zero harm and practice, and a failure to connect with ordinary workers on site. Some employees take the view that the language of zero harm means that their employers are focusing on mechanistic, rather than humanistic, approaches to health and safety on their work sites. It can also again jar with reality – accidents do happen, and so this discourse is simply one that will not "fit" with workers' lived experiences (Sherratt, 2014), no matter how it is presented to them.

- *Business as usual*: When contractors adopt zero harm programmes for the wrong reasons, there is a tendency to continue with the same old work practices, where the adoption of zero harm exists only on paper. If organisations fail to alter their safety strategy to achieve the stated vision, the workforce will realise that nothing has changed, and as a result, their commitment to zero harm is also reduced. Continuation of a "business as usual" approach to health and safety, despite the adoption of zero harm, is tantamount to lip service. As evidenced by the research of Young (2012) into the successful implementation of a zero accident vision in the New Zealand aluminium smelters industry, the need for reforms in work practices and processes meant that a significant contribution had been made towards reducing lost time and the number of injuries.

- *Unintended consequences*: It is worth noting that although administrative accountability within organisations has brought gains, as is demonstrated by some reductions in harm, it also appears to be generating inability to predict unexpected events, and to be creating new safety problems (Dekker, 2014; Hale et al., 2013). For example, when zero harm strategies are undertaken, a substantial number of less visible forms of harm can be ignored. Stressors, for instance, may not be countable, and may, therefore, be ignored until a major event occurs. The administrative accountability of zero harm may consist in counting absenteeism, bruising, cuts, and hospitalisation, without counting the harm associated with fatigue, poor mental health, social dysfunction, and psychosocial illness, etc. Seen through this lens, zero harm is therefore hardly zero harm but has instead become selective zero harm.

Among the criticisms briefly explored in this section, it is perhaps the criticism of the "meaning" of zero harm that dominates. How zero harm is understood and talked about, and consequently accepted by an organisation, is critical. This was demonstrated by the work of Sherratt (2014), in which she examined the practical realities of safety-related "zero targets" in the UK. Research into five large construction contractors operating a variety of zero harm safety programmes determined that "zero" was viewed as both a philosophy *and* a target, with different interpretations in practice between corporate management, site supervision, and the workforce. Such multiple interpretations appear to be challenging the ability of zero harm to support an effective change in practice.

The "meaning" of zero harm has also made commentators see zero harm only as a "target" that sets organisations on a quest for the absolute, something that is not only difficult but likely impossible, for fallible humans to achieve. Such a conclusion can again only come from the accountability perception and consequent reasoning about what zero harm means in practice. Indeed, it has been stressed by Zwetsloot et al. (2013) that the drive to achieve fatality- and injury-free workplaces in industrial systems should not be confused with an accountable zero harm *vision*, which is much more aligned to safety control mechanisms and bringing change to practice. It is this apparent confusion that is pervading both empirical and industry commentaries, which supports the criticism that zero harm is a goal that is both unrealistic and impracticable. Clarity about what zero harm means in an organisation could be achieved in a workplace where people believe that injuries and accidents are preventable, and where they see the concept of zero harm as a fundamental value guiding the operations of the enterprise. If such beliefs and values are embedded in the apparent safety culture of the organisation, the organisation will be able to drive the zero harm vision.

The zero harm vision and safety culture

In contrast to the criticisms of zero harm, there is a growing body of cases demonstrating its success. Case studies of organisations that have recorded success *always* indicate the need to execute credible safety initiatives, activities and other safety programme changes to support the vision of zero harm. The work of the Zero Harm at Work Leadership Program (ZHWLP) in Australia attests to this notion. The ZHWLP involves the members of more than 250 organisations from across different sectors of the Australian economy, all of whom have one common idea: the use of a safety culture to reduce and eliminate the number and severity of incidents of harm in the workplace. (More information on this

programme is available from www.worksafe.qld.gov.au/construction/articles/
zero-harm-at-work-leadership-program).

Here, zero harm is a vision, a need for a positive safety culture, where known
risks are controlled, and unforeseen risks and hazards are addressed in a timely
and effective manner (Welck and Sutcliffe, 2001). Such a safety culture approach
engenders a continuous will to improve health and safety in the workplace. For
example, members of the ZHWLP believe that all injuries are preventable
because zero harm is a way of working towards continuous improvement, yet the
firms do not promote the vision with a zero target mentality. Safety targets are
not destinations, but milestones to measure progress in an improvement journey.
The zero harm vision is seen as a way of thinking, which is influencing safety
behaviours in the workplace, so when setbacks do occur, efforts are made to get
to the main issues and resolve them in practice, rather than playing a numbers
game or the blame game.

The approach used by these firms is akin to Kantianism, a branch of philosophy that follows the works of Immanuel Kant, who believed and promoted the
notion that rational beings have dignity and should be respected – we should
treat people as an end in themselves, not as a means to other things (Kant, 1993,
2012, 2013; Kant and Friedman, 2004; Kant and Guyer, 1998). Kantianism is
about rational morality and freedom, which does not allow blame games, or other
things deemed to be offensive to the dignity of people. It constitutes a significant
aspect of a generative culture, and this can, in turn, be applied to the promotion
of continuous improvement of safety in the workplace. For example, Zwetsloot
et al. (2013: 45) noted that:

> ZAV [zero accident vision] is not a risk control strategy, but a *safety commit-
> ment* strategy. It is an ambition the company commits itself to to achieve
> better safety performance. In the accepted 'control' vision to safety (which
> is dominant in safety research), commitment (and leadership) is frequently
> identified as a major prerequisite; but it is not regarded as the starting point
> for safety improvement. In company practice, commitment to safety creates
> better risk reduction and control strategies, not the other way around […]
> ZAV provides a clear safety message from the top management within and
> outside a company, and it can boost the safety culture.

The above quote resonates with the efficacy of zero harm in safety research and
perhaps goes some way to silencing its critics. The zero harm vision is enhanced
and becomes more effective when *philosophy* is its driver, through thought, reason,
thinking, knowledge, ideology, beliefs, convictions, tenets, values, principles, attitudes, and viewpoints. Given that a safety culture can be defined as the product of
individual and group values, attitudes, competencies, and patterns of behaviour
that determine the commitment to, and the style and efficacy of, an organisation's
safety programme (Health and Safety Commission, 1993, cited in Glendon and
Stanton, 2000: 201), it becomes clear that a positive safety culture is crucial to the
success of any zero harm vision, including the need to involve collective rather

than individual efforts to make the idea work, because both technical and social innovations are required to solve known and unknown safety problems.

Historically, the concept of a safety culture is often cited in association with accident causation models (Peckitt et al., 2004), initially developed from attempts to explain the Chernobyl nuclear disaster of 1986 (Ostrom et al., 1993; Glendon and Stanton, 2000). Safety cultures are concerned with shared attitudes, behaviours, beliefs, norms, practices, systems and values necessary for effective safety controls (Guldenmund, 2000; Glendon and Stanton, 2000; Peckitt et al., 2004). The constituent beliefs and attitudes manifested in safety cultures determine safety performance, and the typologies of safety cultures show that regulation, engineering (safety control), procedure, and behaviour are their main categories (Shillito, cited in Peckitt et al., 2004). Three dimensions of safety cultures, namely behavioural, corporate and psychological dimensions, have also been proposed (Zou and Sunindijo, 2010). Drawing on these categories/dimensions, it is behaviour that gives rise to the empowerment of workers, which is a highly beneficial aspect of safety cultures, as evidenced by research demonstrating that delegation of safety activities to the workforce is consistently correlated with lower injury rates (Törner and Pousette, 2009). The behavioural category of safety cultures includes policies, goals, objectives, procedures, manuals, records, and audits that are used as tools to aid improved performance. In essence, organisations can reduce harm in the workplace because the behavioural category of safety cultures is an enabler of safety compliance (Clarke, 2003; Hafey, 2015), and it is, therefore, able to support the development of practices that, in turn, support the zero harm vision in practice.

Reasons to move towards the zero harm vision in construction

The construction industry worldwide is in need of proactive and significant change in the management of the health, safety, and well-being of its workers. This is particularly critical in developing countries, where the ILO is recording a significant number of fatalities and injuries in its worker database. For example, the safety statistics from Malaysia and South Africa show that biological, chemical and physical hazards are killing and maiming people in the industry – multiple categories of falls and struck-by accidents have been recorded (Emuze and Smallwood, 2015). Clearly, zero harm is not a reality in construction, and, as such, it is perhaps easy to think with scepticism about it. Even in developed countries, where construction harms fewer workers, and where many construction organisations have adopted zero harm in a number of different forms, with the aim of protecting their workers from work-related harm (Wilkens, 2011), the potential remains for incoherence and inconsistency in the understanding of what zero actually means (Sherratt, 2014).

In the introduction to this chapter, the question was asked "How can we make zero harm work for the construction industry?" and this chapter has since offered some critical insights and considerations that can facilitate both the implementation and even perhaps the ultimate attainment of zero harm. Instead of having an inflexible zero target, this chapter proposes that the construction industry seeks to realise a zero harm vision, a philosophical approach that closely aligns with an organisational safety culture, to bring collaboration to the resolution of industry problems. The zero harm vision has the potential to bring change to health and safety programmes used in construction practice, by moving away from a "business as usual" approach, and instead triggering innovation (Zwetsloot et al., 2017) and positive developments in "the way we do things around here", thereby promoting a positive safety culture. Avoiding a "straitjacket" approach, as found with bureaucratic administrative accountability, the zero harm vision should be executed as an aspiration that people should strive towards for the benefit of all. Contractors should adopt it as the foundational philosophy for construction health, safety, and well-being. Instead of using the terms that lay so much emphasis on numbers and "zero", continuous-improvement terms, such as "no harm to anyone at work", or similar terms, could be used to drive the philosophy. When the term "zero" does feature in the vision, there is a need to educate everyone that it is a worthy journey, rather than a hard-and-fast target. The use of Kantianism, for example, as an underlying philosophy for zero harm could assist in mitigating "numbers games", "slogans", "language effects", and skewed interpretations of the actual intentions of the vision.

The vision should not be sold to workers as a "magic bullet" for banishing all forms of harm in the workplace. With the right thinking, it should be introduced and implemented as a way of helping organisations identify and address both hazards and risks in construction work and its workplaces. Gradual reduction and ultimate elimination of harm should be the aim of the vision. As argued by Sherratt (2014), simply setting the target without additional change will not eliminate high numbers of fatalities and injuries in the industry. Instead, the industry should look beyond the statistics and ask why practice immediately reacts to ideas of zero as a target that is simply unattainable. In changing the way zero works, in creating zero harm vision, improvements can be achieved. For example, Wright (2012) reports that zero fatalities were recorded in the London 2012 Olympics project, which adopted a "zero tolerance" approach to unhealthy and unsafe construction practices. This major project involved over 12,000 people, who worked for 80 million person hours in five years, without a single fatality, despite the enormous complexity and logistical requirements of the project. Zero tolerance is arguably a zero harm vision, not a target, and the reality of the approach was the zero-fatality safety record of this significant major construction project.

The zero harm vision in practice

For the zero harm vision to work for the construction industry, it must be placed within a generative and positive safety culture. Although the zero harm vision can be defined as a collection of individual and group beliefs, norms, attitudes, and technical practices that are concerned with minimising exposure of workers and the general public to risks and hazards in the construction workplace (Zou, 2011), it is also arguable that a positive safety culture is also something that remains elusive in construction industry practice. The key components of a positive safety culture include attitudes, beliefs, flexibility, empowerment, learning, the management system, motivation, and values, as illustrated in Table 13.1.

The elements shown in Table 13.1 are all evident in contracting firms that have reported the successful use of the zero harm vision. Taking a cue from the work of Zou (2011), the following examples from four major contracting organisations in the construction industry can be shared:

- *Zero accident philosophy*: Bechtel, a global engineering, construction and project management organisation with more than 50,000 workers in over 50 countries, operates a "zero-accident philosophy". The commitment strategy of the organisation involves all organisational strata, and every individual worker reportedly has the authority to stop work when it becomes or is unsafe. Bechtel promotes the ideology that every accident and injury is preventable. Rather than a safety control paradigm, a philosophical approach to the vision is adopted, positioning safety as a core value in any process and workplace. Because values do not change as priorities do, the organisation ensures that everyone connected to its business sees safety as a core value.

Table 13.1 The characteristics of a positive safety culture

Constituent	Characteristics
Values and beliefs	The values and beliefs found in workplace safety are deeply and widely shared within an organisation.
Attitudes and beliefs	The workers in an organisation have particular patterns of attitudes and beliefs regarding safety practices.
Flexibility	The workers in an organisation are alert to unexpected changes, and they ask for help when they encounter an unfamiliar hazard. They are proactive and will stop work if required.
Empowerment	The workers in an organisation will seek and use available information that improves health, safety, and well-being in the firm.
Management system	The organisation exposes its workers to a safety management system, which is applied in practice and subjected to periodic reviews.
Motivation	The organisation motivates its workers to call attention to safety problems, by giving them recognition and rewards. These workers will think out of the box, to find solutions to identified risks and hazards.
Learning	The organisation deploys systemic mechanisms to collect and process safety-related information. Collection and analysis of such information are done to engender continuous learning of how to work more safely in the firm.

Source: Adapted from Ostrom et al., 1993

- *Zero incidents programme*: Fluor, an integrated engineering, procurement, fabrication, construction, maintenance and project management organisation with over 35,000 workers, operates a "zero incidents" programme. The corporate value of the firm is motivated by the belief that no task is worth a loss of life or injury to the firm's workers. Consequently, health, safety, and well-being matters can be carefully managed. Fluor reportedly specifies responsibilities for management, it promotes accountability, it provides rewards, it provides orientation and training for workers, it establishes a safety task assignment process, and it enables performance measurement and control. The system also allows sharing of collective data from lessons learned that benefit knowledge management efforts.
- *Zero harm*: John Holland Group, a contractor with more than 2,000 workers operating in Australia, has adopted a "zero harm" approach within its safety management system (SMS). The firm describes its zero harm vision for safety on construction sites as sending its people home in the same condition in which they come to work. A notable target within the programme is efforts devoted to a collective endorsement of the vision by all members of the firm's supply chain, in addition to its workers. To implement all the initiatives within the vision, the firm focuses on attitudes and beliefs that lead to safe behaviour.
- *Target Zero*: Carillion, an integrated support services business employing over 46,000 people in the UK, Canada, and the Middle East, operates a programme of "One Culture: Target Zero". Employing both policies and systems to manage health and safety, Carillion is also applying 14 leading and lagging indicators within rigorous risk frameworks, to monitor and improve performance. By combining communications programmes and targeted training initiatives, Carillion is reportedly working hard to achieve the following target zero goal: zero accidents, and zero ill health.

The above case studies adopt a philosophical approach to the concept of zero harm vision. A summary of the lessons learned from such case studies, including those documented in Zou (2011), shows the importance of human factors (attitudes, beliefs, values, mindsets, behaviour, etc.) and a philosophical approach to the implementation and promotion of zero harm visions. It is also worth noting that the different names given to the programmes did not seem to hinder implementation of the programmes, insofar as shaping the belief and value that all incidents and injuries are preventable and unacceptable to management, workers, and the general public were concerned. The zero harm visions demonstrate the crucial role of commitment strategy and leadership from management, which is continually the shaping safety culture in the organisations and their supply chains.

Through the work of organisations such as Safety Action in Australia (Safety Action, n.d.), the Vision Zero Movement in Singapore, and similar organisations worldwide, it can clearly be seen that the philosophical approach to zero harm vision serves the interests of people in construction, instead of the concept simply being used as safety propaganda. However, the vision will remain a pipe dream if it does not have reliable and effective programmes for its execution, where such programmes should continually address hazards and risks in the workplace. When an organisation makes the strategic decision to adopt the vision of zero harm, it is prudent to start with a clear problem that requires urgent attention. After that, the implementation should be clearly communicated as an aspiration that will seek to reveal and address all known and unknown problems progressively. As shown in the case studies above, the use of appropriate and engaging language around the vision is important, starting with the term to be used to share the vision. When workers and all concerned parties can identify with the true meaning of the vision to reduce and eliminate safety problems through credible evidence in their workplace, the vision can then become effective. All these lessons learned, including the accountability and behaviour dimensions, resonate with the common key principles of zero harm vision, as found in Australia, Canada, Finland, Germany, New Zealand, Sweden, the UK, and the United States of America (USA), and as outlined in Table 13.2.

Table 13.2 Elements of essential zero harm principles in practice

Element	Description
Philosophy	Zero harm vision should be a philosophy, not just a slogan.
Aspiration	Clearly communicate that the zero harm vision is an aspiration, not a fixed destination or target.
Programmes	Reliable programmes should be executed as part of the zero harm vision.
Execution	Programmes executed as part of the zero harm vision should progressively address hazards and risks in the workplace.
Evidence	Programmes executed as part of the zero harm vision should provide visible evidence of practical change and improvements in health and safety in the workplace.
Endorsement	Evidence of improvement in health and safety due to the execution of the zero harm vision should be recognised and accepted by both management and workers.
Target problem	Start the zero harm vision with a manifest problem that requires urgent attention in the workplace.
Start small – think big	Start the zero harm vision with smaller, clearly achievable goals, so that tangible progress can be observed and communicated.
Long-term	Ensure that the zero harm vision is a multi-year vision that allows incremental progress.
Transparency	Map out the plan for the roll-out of the vision, so that everyone can track its progress.
Language	Use language that will ensure that the correct perceptions of the vision evolve in the workplace.

Summary

The concept of zero harm can readily benefit the construction industry if it is applied with the right intentions. However, use of the term "zero harm" alone will not bring any real benefits; rather, it is the initiatives that are associated with the intentions of the vision, and the change that the vision can inspire and bring about, that form the core cluster of benefits. For example, firms that fail to implement credible initiatives or actions as part of the vision will also fail to engender continuous safety improvement. It is not enough to simply adopt the vision and display it on signboards around the work site and print it on workers' uniforms and hard hats and company vehicles. Firms that fail to implement concrete actions as part of their adoption of zero harm represent case examples for scepticism and resistance. Paying lip service to the zero harm vision, either through a lack of real change, through deliberately using the vision as a marketing tool or public relations slogan, or through focusing on "number-counting", will not make the vision work for the construction industry. The language used to promote the vision, and the discourse that it should subsequently develop is also crucial to ensure that the meaning of the vision is shared and communicated effectively to all those concerned.

Case studies have shown that major contractors in developed countries recognise the importance of zero harm. However, they must be cautious of simply rebranding "health and safety business as usual". The zero harm vision needs to inspire real change in practice, to support the development of a generative positive safety culture, and to communicate this and seek engagement on the site, throughout the organisation, and all along its supply chains. Only then can the real potential for the zero harm vision to reduce harm to people in construction be truly realised.

Notes

Many of the ideas presented in support of the zero harm vision rely on arguments made in the following texts:

- Emuze, F.A. and Smallwood, J.J. (2015). Towards zero construction motor vehicle accidents in South Africa. In: *Proceedings of the CIB W099 International Health and Safety Conference*, 9–11 September, Belfast, UK, pp. 72–81.
- Smallwood, J.J., and Emuze, F.A. (2015). Towards zero construction fatalities, injuries, and diseases. In: *Proceedings of the 8th International Structural Engineering and Construction Conference*, 23–28 November, Sydney, Australia, pp. 1279–1284.
- Zero harm: Infallible policy or ineffectual theory? (2013). *OHS Professional*. Available online at https://sia.org.au/downloads/public ations/ohs_professional_magazine/ohs.march2013.pdf (Accessed 21 October 2016).

References

BP (British Petroleum) (2010). *Deepwater Horizon* accident investigation report. Available online at www.bp.com/content/dam/bp/pdf/sustainability/issue-reports/Deepwater_Horizon_Accident_Investigation_Report.pdf (Accessed 5 April 2016).

Clarke, S. (2003). The contemporary workforce: Implications for organisational safety culture. *Personnel Review*, 32(1), 40–57.

CWPR: The Center for Construction Research and Training (2013) *The construction chart book: The US construction industry and its workers* (5th edn). Silver Spring, MD: CWPR. Page 37. Available online at: www.cpwr.com/sites/default/files/publications/CB%20page%2037.pdf (Accessed 29 March 2017).

Dekker, S.W. (2014). The bureaucratization of safety. *Safety Science*, 70: 348–357.

Dekker, S.W., Long, R. and Wybo, J-L. (2015). Zero vision and a Western salvation narrative. *Safety Science*, 88(2): 219–223.

Emuze, F.A. and Smallwood, J.J. (2015). Towards zero construction motor vehicle accidents in South Africa. In: *Proceedings of the CIB W099 International Health and Safety Conference*, 9–11 September, Belfast, UK, pp. 72–81.

French, S., Bedford, T., Pollard, S.J.T. and Soane, E. (2011). Human reliability analysis: A critique and review for managers. *Safety Science*, 49(6): 753–763.

Glendon, A.I., and Stanton, N.A. (2000). Perspectives on safety culture. *Safety Science*, 34(1): 193–214.

Guldenmund, F.W. (2000). The nature of safety culture: A review of theory and research. *Safety Science*, 34(1): 215–257.

Hafey, R.B. (2015). *Lean safety Gemba walks: A methodology for workforce engagement and culture change*. Boca Raton, FL: CRC Press.

Hale, A.R., Borys, D., Adams, M., (2013). Safety regulation: The lessons of workplace safety rule management for managing the regulatory burden. *Safety Science*, 71: 112–122.

Hardin, G. (2009). The tragedy of the commons. *Journal of Natural Resources Policy Research*, 1(3): 243–253.

Haslam, R.A., Hide, S.A., Gibb, A.G., Gyi, D.E., Pavitt, T., Atkinson, S., and Duff, A.R. (2005). Contributing factors in construction accidents. *Applied Ergonomics*, 36(4): 401–415.

Hollnagel, E. (2011). Prologue: The scope of resilience engineering. In: E. Hollnagel, J. Pariès, D.D. Woods and J. Wreathall (eds), *Resilience engineering in practice: A guidebook*. Farnham, UK: Ashgate.

Hollnagel, E. and Woods, D.D. (2006). Epilogue: Resilience engineering precepts. *Resilience engineering: Concepts and precepts*. Aldershot, UK: Ashgate. pp. 347–358.

Kant, I. (1993). *Grounding for the metaphysics of morals: On a supposed right to lie because of philanthropic concerns*. Indianapolis, IN: Hackett Publishing Company.

Kant, I. (2012). *Fundamental principles of the metaphysics of morals*. Mineola, NY: Courier Corporation.

Kant, I. (2013). *An answer to the question: 'What is enlightenment?'* London: Penguin UK.

Kant, I. and Friedman, M. (2004). *Kant: Metaphysical foundations of natural science*. Cambridge: Cambridge University Press.

Kant, I. and Guyer, P. (1998). *Critique of pure reason*. Cambridge: Cambridge University Press.

Kheni, N.A., Dainty, A.R.J. and Gibb, A. (2008). Health and safety management in developing countries: A study of construction SMEs in Ghana. *Construction Management and Economics*, 26(11): 1159–1169.

Long, R. (2012). The zero aspiration, the maintenance of a dangerous idea. Available online at www.safetyrisk.net/wp-content/uploads/2014/04/The-Zero-Aspiration.pdf (Accessed 21 October 2016).

Manuele, F.A. (2014). Incident investigation: Our methods are flawed. *Professional Safety*, 59(10): 34–43.

Ostrom, L., Wilhelmsen, C. and Kaplan, B. (1993). Assessing safety culture. *Nuclear Safety*, 34(2): 163–172.

Peckitt, S.J., Glendon, A.I. and Booth, R.T. (2004). Societal influences on safety culture in the construction industry. In: S. Rowlinson (ed.), *Construction safety management systems*. London: Spon Press. pp. 14–48.

Rawlinson, F. and Farrell, P. (2010). UK construction industry site health and safety management: An examination of promotional web material as an indicator of current direction. *Construction Innovation*, 10(4): 435–446.

Rosness, R., Blakstad, H.C., Forseth, U., Dahle, I.B. and Wiig, S. (2012). Environmental conditions for safety work–theoretical foundations. *Safety Science*, 50(10): 1967–1976.

Safety Action (n.d.) The safety action team. Available online at www.safetyaction.com.au/about/the-team/ (Accessed 21 October 2016).

Sherratt, F. (2014). Exploring 'zero target' safety programmes in the UK construction industry. *Construction Management and Economics*, 32(7–8): 737–748.

Sherratt, F. (2016). *Unpacking construction site safety*. Chichester, UK: Wiley.

Snowden, D. (2000). Cynefin: A sense of time and space, the social ecology of knowledge management. In: C. Despres and D. Chauvel (eds), *Knowledge horizons: The present and the promise of knowledge management*. Oxford: Butterworth Heinemann.

Snowden, D. (2002). Complex acts of knowing: Paradox and descriptive self-awareness. *Journal of Knowledge Management*, 6(2): 100–111.

Steen, R. and Aven, T. (2011). A risk perspective suitable for resilience engineering. *Safety Science*, 49(2): 292–297.

Swuste, P. (2012). Editorial – WOS2010, on the road to vision zero? *Safety Science*, 50: 1939–1940.

Törner, M. and Pousette, A. (2009). Safety in construction – a comprehensive description of the characteristics of high safety standards in construction work, from the combined perspective of supervisors and experienced workers. *Journal of Safety Research*, 40(6): 399–409.

Welck, K.E. and Sutcliffe, K.M. (2001). *Managing the unexpected: Assuring high performance in an age of complexity*. San Francisco: Jossey-Bass.

Wilkens, J.R. (2011). Construction workers' perceptions of health and safety training programmes. *Construction Management and Economics*, 29(10): 1017–1026.

Wright, E. (2012). Olympic health and safety: Record breakers. *Building*, 1 June.

Young, S. (2014). From zero to hero. A case study of industrial injury reduction: New Zealand Aluminium Smelters Limited. *Safety Science*, 64: 99–108.

Zou, P.X. (2011). Fostering a strong construction safety culture. *Leadership and Management in Engineering*, 11(1): 11-22.

Zou, P.X.W. and Sunindijo, R.Y. (2010). Construction safety culture: A revised framework. In: *Proceedings of Chinese Research Insitute of Construction Management 2010 International Research Symposium (CD ROM)*, Johor, Malaysia.

Zwetsloot, G.I., Aaltonen, M., Wybo, J-L., Saari, J., Kines, P. and De Beeck, R.O. (2013). The case for research into the zero accident vision. *Safety Science*, 58: 41-48.

Zwetsloot, G.I., Kines, P., Wybo, J-L., Ruotsala, R., Drupsteen, L. and Bezemer, R.A. (2017). Zero Accident Vision based strategies in organisations: Innovative perspectives. *Safety Science*, 91: 260-268.

14 A Kantian approach to respect-for-persons in construction site work

Fidelis Emuze and John Smallwood

Dare to think/Dare to be wise

(Immanuel Kant)

Background

'Respect for people' is a principle regarding the role of people in construction where they undertake work on sites. To respect construction people means to regard them highly by valuing their unique contributions to the built environment and society. The plight of people in construction is well documented in both print and electronic media. The difficulties are evident in any database that hosts construction-related information. Problems plaguing people in construction are interwoven with working conditions that impact upon their health, safety, and well-being. These problems occur either when people are active in the industry or when they exit the industry (either prematurely or after retirement), or both. People issues in construction cannot be ignored, because people undertake the physical and non-physical work demanded by clients, they operate the plant and equipment, and they supervise and manage the construction process to deliver projects in a socio-technical industry. These roles of people in construction make them the most valuable resource in an industry where research and development (R&D) and the adoption of technology lag behind those of manufacturing and other major industrial sectors (Forbes & Ahmed, 2010). Commitment to people that is evident in respect and decent workplaces is thus crucial for the progress of the construction industry because of their roles in the process.

As opposed to 'respect for people' as a term, this chapter adopts the original terminology, which is 'respect-for-persons' (RfP). The RfP phrase entails a sense of deferring to others by considering their interests and looking out for their well-being, even if doing so places certain constraints on oneself (Lysaught, 2004). The notion of respect is necessary for the RfP concept. The RfP explained in this chapter refers only to that respect that is owed equally to all persons by their status as human beings with moral personalities. Any person, good or bad, is entitled to such respect by being a person, an honest character for whom it is good to act upon and satisfy required interests, aims, purposes and objectives (Hudson, 1980). The chapter sets out to convince readers that Kant's moral philosophy has significant implications for construction practice. When the three formulations of the categorical imperative explained in this chapter are logically considered, they give guidelines to construction regarding negative and positive ideas that impact upon people in the industry: individuals in construction should be treated as ends, never just as means. The use of the formulations would prohibit actions that eventuate in coercion and deception, apart from encouraging a more decent workplace that is healthy and safe for the promotion of meaningful work.

Origins of respect-for-persons

RfP is situated within the Kantian moral theory framework. Immanuel Kant (1724–1804) was born in Königsberg in East Prussia, which is located in a Russian territory that is now called Kaliningrad. Kant is renowned for his writings on ethical theory. The main elements for positive social interaction found in Kant's writings include (a) the roots of respect that are directed at persons as rational beings; (b) respect for the moral law that is a product of reason guided by judgement; and (c) the idea of ethical conduct in good will, which qualifies respect in terms of humanity (Wawrytko, 1982).

The origins of RfP express a minimum respect for persons and highlight the following Kantian principle:

> Act in such a way that you always treat humanity, whether in your person or the person of any other, never simply as a means, but always at the same time as an end.
>
> (Teuber, 1983: 370)

What does Kant mean by this idea? According to Kant's writings, in the realm of ends, everything has either a price or a dignity. Whatever has a price can be replaced by something else as its equivalent; while whatever is above price, and thus has no equivalent, has a dignity (Kant & Friedman, 2004). In other words, people are owed respect by their dignity that has no price tag. The big picture of RfP is that people are not respected because they happen to measure up to some standards (real or imaginary): rather, people are recognised for being who they are. This picture is important because when respect is based on an individual characteristic or attribute, the worth ascribed to such persons will no longer be incomparable (there is a price tag involved in such considerations). Rational nature and reason make every one unique, and as long as a person is both, a price tag is not applicable. Rationality is the common basis for human interaction, and reason can identify the content and method of approach to morality (Wawrytko, 1982). Kant employs RfP and deems people to be worthy of respect regarding an *a priori* concept of reason, which comes from rationality in people (Wawrytko, 1982).

Similar to a work of art, respect in the Kantian philosophy is given regardless of considerations for the usefulness of its objects (Kant & Friedman, 2004; Teuber, 1983). In this philosophy, respect is not for the person *per se*, but the moral law that is practically alive to show the absoluteness of worth. That is why Teuber (1983: 375) says:

> The person is an end-in-himself, but only insofar as he has within him something that is itself an end-in-itself. An individual is only of incomparable worth insofar as he has within him something of unique worth. Each person is unique and irreplaceable only insofar as each person has within him something that is unique and for which there can be no substitute. What is unique and irreplaceable in the person? Well, it is the moral law, of course; and instead of respect for the individual as such, we have respect for the moral law as such.

So moral law cum conduct demands respectful attitudes and actions that underpin ethics and social interactions (Wawrytko, 1982). There is therefore only one thing of absolute value, and that is the moral law that has reason and rational nature (Kant, 1949; Kant & Guyer, 1998). RfP thus dictates the subordination of personal will to the moral law under the discipline of reason. The subordination to the moral law is crucial as when placed under the discipline of reason, morality tends to be more reliable because individual mood swings will not dictate decisions and actions:

> Morality exists, in part, to remind us when we are inclined in one direction that, perhaps, it would be better if we went in some other.
>
> (Teuber, 1983: 385)

As such, a genuinely moral person acts according to respect for moral law and not for some good that will come from such action and the moral law is worthy of respect because it helps people to rise above the natural order of cause and effect regarding doing what is universally accepted as right or correct. By subjecting themselves to the dictates of moral law, people can slowly minimise the personal nature of doing things.

With recognition respect, appraisal respect, evaluative respect, directive respect, institutional respect, obstacle respect, identification respect and so on, the philosophy of ethics has sought to give various connotations and descriptions to the RfP concept (Cranor, 1983; Darwall, 1977; Hudson, 1980). Recognition respect acknowledges the value in an object, while Appraisal respect is about some judgment on how great the value in the object is (Darwall, 1977). Identification respect assumes that moral agents can identify with all other human beings and its takes the moral perspective (Cranor, 1983). Respect, therefore, comes in a variety of forms. For example, people are recognized for uprightness or exceptional achievements, or both. However, most individuals accord respect to people merely because they are people and such respect is of keen interest to moral theory (Buss, 1999). Seen as an alternative to utilitarianism, Kant's principles of RfP have become mainstream on account of morality and how RfP bears on duties, rights, autonomy, equity and justice (Hudson,

1980). The broad adoption and adaptation of this concept of RfP are based on Kant's assertion that it applies to everyone, regardless of social position, real accomplishment, prestige, and personal merit. This notion of RfP is what is needed in an industry where social interaction involves all strata of society, especially on construction project site work.

The philosophy of ethics provides a common basis for social interaction. The work of Kant is mainly about the concept of respect as the backbone of ethical systems that are essential for ethical practice (Wawrytko, 1982). The most critical concept in ethical practice is the idea of 'goodness', which is deemed to be enduring and absolute (Gaita, 2004). Ethical behaviour is motivated by respect for the moral law, which seeks to promote the highest good as a cooperative effort in society (Wawrytko, 1982). In this regard, the 'good' is not so much of a pursuit. Instead, it is the way individual and collective pursuits are judged (Gaita, 2004). The RfP notion from Gaita (2004) enriches the work of Kant by emphasising that people are owed genuine respect for what they *are*. The notion reiterates the work of Kant with the view that all persons are inimitable. The vital point is that wherever people find themselves, they should always act within the context of morality so that unconditional respect and technical debates do not conflict with each other (Gray & Stofberg, 2000).

The interaction between morality and moral qualities can be illustrated with the interplay between artisans and the furniture made by them. This communication from the writings of Gaita (2004) flagged the fact that carpenters tend to:

> ...care for the wood in ways which are not reducible to its functional properties, although it cannot be entirely independent of functional considerations. Craftsmen and all those who understand themselves regarding having a vocation, 'are engaged in a limitless process of self-exploration through an exploration of what they do'.
>
> (Gaita, 2004: 87)

The interaction shows that the work that artisans do cannot be reduced to mere functionality. Taking a cue from Gray and Stofberg (2000), Gaita's notion of RfP may be applied to construction for the following reasons:

- People are not respected for being either perfect examples of goodness or striving towards goodness. But people are respected because of the humanity in them (their inherent humanity).
- What it is to be a human being comprises being an irreplaceable individual who has a unique role within complex relations in a project.
- RfP forbids seeing people as people with 'lesser status'. Disrespect to people limits their will.
- RfP demands the integration of thought, feeling, and knowledge of the contextual networks in an environment. For instance, being a construction professional does not imply lack of concern for the well-being of workers when the situation demands it.
- RfP requires knowing oneself regarding integrity, which is essential for uniqueness in every situation.

A Kantian sketch of respect-for-persons for construction

The literature shows that Kant's philosophical and ethical theories are met to guide society regarding moral intentions and duties. Duty about moral law should be topmost in the minds of human beings so that people do not promote individual inclination to do as they wish and please (Kant, 1949). People should avoid conflict between certain personal ends and the moral law of duty (Kant, 1974). Thus, Kant's 'categorical imperative' (1785/1993) is intended to be the moral norm that society should follow by making decisions of which the outcomes are accepted and acceptable by everyone (Place, 2010).

Kant argued that good persons should do their duty and described these duties as categorical. To show that people are capable of moral worth, Kant says that it is sufficient to show that they are rational beings (Teuber, 1983). So Kant referred to the fundamental principle of ethics as the categorical imperative (Bowie, 1998). Kant contends that reason is the foundation of the categorical imperative (Johnson, 1974; O'Neill, 1989; Westphal, 1995; Williams, 1992). Reason is the source of categorical imperative because when reason is on the side of an action, then a rational agent (human being) can endorse the action; and if an intelligent agent can approve an action, then the acceptability of the action does not depend on feeling or individual inclinations (Buss, 1999). For this chapter, three formulations of the categorical imperative have been used to sketch Kantian RfP that is tenable for the construction industry. These formulations include the following (Bowie, 2002):

- Act only on principles that you can promote to be universal laws of nature (formulation 1).
- Always treat the humanity in a person as an end, and never as a means (formulation 2).
- Act as if you were a member of an ideal kingdom of ends in which you were subject and sovereign simultaneously (formulation 3).

According to Bowie (2002), Kant's ethics is about duty, and the deontology is rule-based, and it contends that duty applies in all situations, such as 'Thou shall not kill' in the Bible. So the moral person is the person who acts from the right intentions (Louden, 1986; O'Neill, 1993). RfP principles have thus been used to justify moral rights and duties regarding the rights to which a person is entitled out of respect and the duty owed to others out of respect for them (Cranor, 1975). The fundamental principle of Kant's ethics, the categorical imperative, is a requirement of reason and is obligatory for all rational human beings (Bowie, 1998, 2002). Similar to the work of Bowie (2002) where the categorical imperative was applied to business ethics, the next three sub-sections of this chapter apply it to the construction context.

The futility of immoral actions – formulation 1

Kant's first formulation of the categorical imperative appears to be assessing whether a proposed action, including actions either on a construction site or in the head office of a construction enterprise, is moral. The formulation functions as a check to determine whether the principles upon which an action is based are morally acceptable. The action can only be embarked upon if the principle on which the action is based passes the categorical imperative evaluation (Bowie, 2002). For instance, a manager who accepts Kantian ethics would evaluate the basis of a construction project decision regarding universal application and acceptability without contradiction. If the decision can be universally applied, then the decision would be morally acceptable (Nash, 1990). Conversely, if the decision cannot be universally applied, then it is morally prohibited (Nash, 1990).

To support Kant's point, an example is inferred directly from the work of Bowie (2002). Theft by workers, managers, and the public is a problem, particularly on a less secured construction site. Suppose that a general worker with limited trade skill is angry at his foreman for some justified reason (illegal dismissal, for example) and considers stealing from the project. Could a principle that permits theft be universalised? Clearly, such a principle could not. Since goods and services are in limited supply and universal shared ownership is difficult, private property ownership is in force. If a principle that permits stealing is universalised, private property ownership is null and void. What is being illustrated is that when people are free to take from each other, then all forms of ownership are in jeopardy. With this example, Kant is arguing that if there is a practice of private property then a principle that permits theft is logically self-contradictory and futile. Thus, if a construction worker who is unfairly fired from a project steals from the project, the act is morally unacceptable.

The illustrated example indicates that making exceptions for oneself outside actions that are underpinned by a principle that can be universalised is immoral and unethical (De George, 2011; Merrill, 2011). In essence, this formulation of the categorical imperative is relevant in the construction industry. If a principle for an action, when universalised, is self-defeating or futile, or both, then the anticipated action and its consequences would not be ethical. Based on Kant's theoretical perspective, when people in construction behave recklessly through individual site practices such as the use of improper personal protective equipment (PPE) or the non-usage of it, they are projecting immoral or unethical characteristics. For example, an exploratory Lesotho study that focuses on PPE pinpoints the reasons for the poor use of PPE through the perceived attitudes of employers in the construction industry (Emuze & Khetheng, 2016). The interviewed workers in the study were unanimous about implementation and enforcement gaps. The study reveals that construction workers in Lesotho may proceed with construction site activities without the required PPE owing to either limited or total lack of requisite site inductions and inadequate H&S site supervision – no H&S officer on site. Regarding the client, the workers say it seems in most cases clients do not take the initiative about H&S issues. According to them, this is why contractors benefit from the workers by not ensuring the right working conditions regarding PPE.

Treating construction workers as persons – formulation 2

Kant's moral theory postulates that human beings have a free will and can act from laws required by reason because they have a dignity that places a value beyond price on them (Kant, 1993, 2012). One human being should not exploit another person (Kant, 1993, 2013). When people do not grant respect to others as rational beings, as Kant contended, people can be manipulated in the process of using them as means instead of as ends (Modarelli, 2006). When people are manipulated as means to get to an end, there is always the possibility of taking unfair advantage of them based on the notion of exploitation, which is bad (Sample, 2003). Exploitation is the opposite of good as it not only takes unfair advantage of people; it could also emerge out of disrespect for the exploited people (Sample, 2003). Fatalities and injuries that befall workers on hazardous construction work sites could be linked to unfair treatment.

Given the imbalance in power relations between workers and their employers (contractors), even when employees can spot and record a potential hazard, they could still become victims. A case in point is the Tongaat Mall building collapse in South Africa where two fatalities and 29 injuries would have been avoided if the developer and contractor had been determined to do so (Emuze, Van Eeden & Geminiani, 2015, 2016). The practice illustrated in the analysed reports by Emuze et al. (2015, 2016) is indicative of deviations from espoused zero harm practice. Emuze et al. (2015) used content analysis as a method to provide possible reasons for the accident. These reasons include poor construction work, inadequate supervision of work, and non-compliance with H&S regulations. Emuze et al. (2016) further show that the developer of the project failed to comply with official orders to stop the project by ignoring court orders to stop construction even six days before the collapse. This disregard for regulations by the developer impacted working conditions and contributed to the recorded injuries and fatalities. In brief, the forensic accident investigation determined that the developer had used cheap materials and cement products, and the need to complete the construction work in time for occupation engendered serious compromises regarding a method of work on a site.

It is, however, pertinent to note that South Africa is not alone regarding the causes of building collapse. Similar factors were reported when buildings collapsed in Malaysia (Aini et al., 2005) and the United States of America (USA) (Levy & Salvadori, 2002). In the South Africa example, the workers realised that some of the work was dangerous and even pointed it out to the contractors. But the imbalance in power relations and high levels of unemployment meant that the workers remained working on the site until the building collapsed. This example is akin to a situation where the employer exploits the workers to achieve an end. In fact, exploitation is the antithesis of meaningful work (Skrivankova, 2010).

In other words, one human being cannot use another only to satisfy his or her interests (Kant, 2012, 2013) without adverse outcomes. For Kant, moral relations between people always imply mutual respect for autonomy (Lysaught, 2004). Mutual respect for freedom is the central tenet from Kant's second formulation of the categorical imperative, which says: "*Always treat the humanity in a person as an end and never as a means*". This formulation can be re-stated as "*Always treat the humanity in a construction worker as an end and never as a means to an end*". The question is what the implications of this second formulation are for construction.

As a caution, it is important to clarify that RfP principles do not forbid profitable economic transactions (Melé, 2009). Rather, this second formulation of the categorical imperative only places some restrictions on the nature of financial transactions (Bowie, 2002). To realise the intentions of Kant with this formulation, Bowie (2002) drew a distinction between negative freedom (freedom from coercion and deception) and positive freedom. Positive freedom is the liberty to develop human capacities (Qizilbash, 1996). The liberty is usually under the umbrella of empowerment in construction. For Kant, freedom means the development of both rational and moral capacities (Kant, 2012). In interacting with others, no one should either reduce or inhibit the development of human capacities (Bowie, 2002). This formulation of Kant's, therefore, requires that people in a work-related relationship should be neither coerced nor deceived. Instead, relationships should contribute to the development of rational and moral capacities rather than inhibit the development of these capabilities as these requirements, if implemented, would change the nature of work relations and practice (Bowie, 2002).

As an illustration, construction workers are always concerned about the massive layoffs created by the volatility of work in the industry, which is rooted in projects that are subject to economic cycles. The concerns mainly surround the morality of layoffs. The first author of this chapter experienced the layoffs in a massive project that was a victim of the 2008 economic meltdown. Although the meltdown had a marginal impact in Africa, redundancy and layoffs were the order of the day in several parts of the world of construction. Fairness in the process is the significant contestation regarding layoffs, especially among women in construction (Lingard & Lin, 2004). It is also difficult to get laid-off talents back into the industry where stability is not assured, and profit-making enjoys *primacy* regarding ideology (Philips, 2003).

It is important to note that layoffs are not always immoral because, supposedly, workers have not been used as mere means to enhance profitability by contractors. Instead of a possible conclusion about layoffs, what would be required from a Kantian viewpoint is an evaluation of the nature of the relationship regarding the possible use of coercion and deception in effecting such layoffs. If coercion and fraud were found to have played a role in the layoffs, then immoral acts would have taken place. The point in this Kantian formulation of RfP is that treating the humanity in a person as an end requires the use of positive actions to help people (Kant, 1993, 2012). Kant's moral philosophy enables businesses to develop a useful definition of meaningful work as it would require firms to provide decent work (Bowie, 1998). Based on the Kantian perspective, meaningful work, according to Bowie (2002):

- is freely chosen and provides prospects for the worker to exercise independence on the job;
- supports the freedom and rationality of people: work that lessens independence or that undermines rationality is immoral;
- provides a salary adequate to exercise liberty and provide for the well-being of workers;
- enables an employee to develop rational capabilities; and
- does not interfere with an employee's moral growth.

These conditions of meaningful work have been amplified as "decent work" requirements by the International Labour Organisation (ILO) (Auer, 2007). Within the construction industry, the concept of decent work is particularly crucial because of working conditions that are dangerous (Abdelhamid & Everett, 2000; Chi, Han & Kim, 2012). The concept of decent work thus mandates that workers should experience their work as meaningful and beneficial (Anker, Chernyshev, Egger, Mehran & Ritter, 2003; Ghai, 2003; Posthuma, 2010).

To close this Kantian formulation, it is important to state that contractors who have adopted the Kantian approach would regard decent work as a moral obligation that is not subject to negotiations or the 'business case' mantra. Respecting people by business case implies the existence of a profitability-related limit which dictates what workers should expect from a firm (Ness, 2010). Such an idea that appears to be advocated in the construction industry is sharply at variance with the origins and intent of RfP. Contractors have to create an organisation (regardless of size regarding a number of employees and business turnover) that upholds the tenets of decent work: it is a moral obligation. This requirement should be viewed as a minimum. The third Kantian formulation of categorical imperative explains what is expected from such firms.

Construction contractors as a moral community – formulation 3

The third formulation promotes the idea of a moral community. Kant's third formulation of the categorical imperative views firms as moral communities. Given that construction companies and other associated organisations in the industry are populated by individuals (professionals, managers, technicians, and craftspersons), structures in these organisations should treat the humanity in people with dignity and respect (as an end instead of means). As a case in point, the rules that govern a contracting firm should be open and acceptable to everyone in the company. A widespread acceptance by concerned people in an enterprise is what enables Kant to say that people are both subject and sovereign on the rules that govern them (Bowie, 2000).

When organisations benefit from society, they have a duty of beneficence to society in return (Beauchamp, 2008). Corporations stand to benefit from society as society protects them by providing the means for enforcing contracts, apart from providing the infrastructure that allows the firm to function. The Kantian moral theory also requires worker participation that is crucial to a safe work site in the construction industry (Choudhry, Fang & Mohamed, 2007; Dedobbeleer & Béland, 1991; Hafey, 2015; Lingard & Rowlinson, 1998; Mohamed, 2002). Certainly, a necessary condition of autonomy is consent given under non-coercive and non-deceptive circumstances. Approval also requires that the individuals in an organisation endorse the rules that govern them. Permission is needed in a team environment where work participation is required to reduce hazards. Teamwork is almost universally praised, and several firms have recognised the concept of culture in construction. Organisational and safety culture has been discussed widely in construction (Cooper, 2000; Glendon & Stanton, 2000; Guldenmund, 2000; Parker, Lawrie & Hudson, 2006).

However, the culture dialogue in construction has to take account of RfP. For example, a way of showing respect is to refrain from some actions that could harm people in construction. What is therefore perceived as being respectful regarding the health, safety, and well-being of people in construction? How respectful are the working conditions found on most construction sites? These reflexive questions are asking actors in the industry to consider how their decisions and actions in an industry that is multicultural in make-up are expressed and viewed. More so, people can hardly be said to respect a person if the individual is treated with contempt regarding all that gives meaning to his/her life and makes him/her the kind of person he/she is (Parekh, 2002). In other words, the Kantian RfP assigns equal human worth to everyone, and its absence shows a failure to admit the equal value of people (Simpson, 1979). Respectful treatment of people in construction is a pressing demand because they are aware of respect and disrespect showed to them; and this awareness prompts corresponding responses from them (Green, 2010). Akin to the implications of the Kantian creed of RfP to the problem of sweatshops in the manufacturing industry (Arnold & Bowie, 2003), this chapter is arguing that duties of construction enterprises should not be neglected on any construction site, especially about physical work on a site. Construction companies have to ensure that regulations in project locations are followed; they have to refrain from coercion and deception in their engagement with workers, and they have to meet the minimum health and safety (H&S) standards while observing the provision of living wages for their employees.

Respect-for-persons discourse in construction

In construction, the RfP reports in the United Kingdom (UK) appear to align with Rawls's principle of utility, which says "Treat people as means and ends" (Ness, 2010). This principle treats people as ends by assigning the same weight to the welfare of each, whereas it treats them as means by allowing the benefits to counterbalance the losses of others, especially when the ones that lose are disadvantaged (Teuber, 1983). The 2000 and 2004 RfP reports in the UK responded to prevailing circumstance of the industry at the time of the publications regarding the shortage of skilled workers, the poor image of the industry, and the need to pre-empt regulations (Respect for People Working Group, 2000; 2004). These two reports covered diversity, working facilities, training and H&S as people issues. While the 2000 report advocated for improvement in investing in people, workforce participation, behavioural issues, and an overarching management framework, the 2004 report focused on enhancing equity and diversity in the workplace, working environments and conditions, H&S, and lifelong learning. In essence, for the industry to respect its people, a transformation is needed in how construction practice promotes diversity, working facilities, training, H&S, and well-being.

The two reports, however, fell short regarding projecting what Kant intended with RfP. As rightly observed by the following analysis, conceptualising people (construction workers) as assets encourages a utility principle that makes it acceptable to treat employees as a means to an end (Ness & Green, 2008). This realisation is exacerbated by the use of the business case argument to drive the need to respect people in construction. Apparently, the industry has failed to adopt the original meaning and use of RfP as advocated by Kant (Bowie, 2000; Bowie, 1998, 2002; Kant, 1949, 1974, 1993, 2012, 2013; Kant & Friedman, 2004; Kant & Guyer, 1998). With leadership, decent work, meaningful work, business ethics, and every other human relational endeavour, RfP is not intended to be anchored on monetary or economic considerations – Kantian philosophy says people have no price tag because they have dignity. For example, the business case arguments of RfP in construction imply that positive changes to working conditions should be engendered mainly regarding contributions to profitability instead of morality (thou shall not kill people) and fairness (protecting people) (Ness & Green, 2008). Protecting workers on construction sites and the general public in project locations is not an economic issue. It is a moral issue as it is wrong to kill and maim people. There can be no justification (moral or economic) for accepting the killing and maiming of individuals as a universal rule.

The lacklustre changes in the working conditions of people in construction, and the dismal conditions of individuals in construction in particular parts of the world make the adoption of the real intent of RfP exigent in the industry (Adsul, Laad, Howal & Chaturvedi, 2011; Ghaemi, 2006; Kuruvila, Dubey & Gahalaut, 2006; Tam, Zeng & Deng, 2004). It is then imperative for actors in the industry to view the Kantian RfP by recognising that:

> To respect a person is simply to respect that person's moral rights. We need not know what moral rights are possessed by persons, as such, to understand the structure of such a theory of respect-for-persons. We need only to know that persons possess some moral rights and that to respect a person is to respect these rights, whatever rights (for example, a right to life) these may be.
>
> (Hudson, 1980: 83)

This quotation touches on the right to life that is guaranteed in the constitution of most nations, although there have been many contestations about the ability of the industry to provide such rights to workers (Dekker, 2014; Dekker, Long & Wybo, 2015; Sherratt, 2014; Young, 2014; Zwetsloot et al., 2013). This right to life should cover everyone involved in construction work where fatalities are evident in developed countries that are allegedly doing better than developing countries. In particular, after the publication of the RfP in construction reports in the UK, the image and performance of the industry have not realised significant transformation, if both empirical and anecdotal evidence are evaluated. The industry in the UK still grapples with fatalities that point to need to up the working conditions in construction so as to eliminate on-site deaths entirely (see Figure 14.1). This realisation endorses the discourse analysis of the RfP in construction reports (Ness, 2010). The question is 'What will it take to ensure that construction workers enjoy the right to life (no harm) as a result of their involvement in the industry?' The answer to this question will entail a journey that must incorporate two RfP moral convictions. These convictions that apply to construction workers recognise autonomy (an individual should be treated as an autonomous agent) and protection (persons with diminished autonomy should be protected) (Lysaught, 2004). Construction workers should be provided with decent working conditions and the ones with compromised status due to economic reasons should see fairness in their treatment. The latter can be done by reducing the gap between the respect shown to 'trade' and 'professional/ management' workers in the industry (Murray, Chan & Tookey, 2001).

In brief, the totality of RfP as a force for positive social interaction is found in the moral role model that is embodied in Kant's goodwill as:

> Nothing in the world - indeed nothing even beyond the world - can be called good without qualification except a goodwill.

> (Kant, 2012: 9)

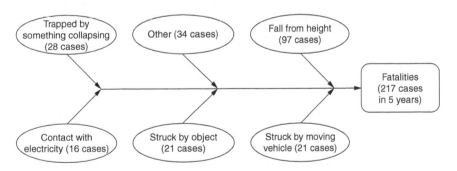

Figure 14.1 Unending work-related fatalities in construction – UK

Source: Health and Safety Executive - H&S in construction sector in GB, 2014/15.
http://www.hse.gov.uk/statistics/adhoc-analysis/lfs-worked-related-rta.htm

Actors in the construction industry should subscribe to the ideal goodwill that is supported by the principle of humanity. RfP goes hand-in-hand with self-respect. The writings of Kant state the inner sense of worthiness and self-esteem should supersede utilitarian concerns (Wawrytko, 1982), which is called 'business case' in the construction industry. When this is accepted, only then can the good be projected regarding people in construction and society as a whole. One should respect the system of moral rights that people enjoy as it represents the fact that each individual's right is entitled to equal consideration (Hudson, 1980). In fact:

> ...to recognise that human beings are equal is to see that respect for persons is warranted and to desire that everyone's needs be equally met. When they are not met, and social arrangements favour some over others the structure of human relations is distorted and one rightly feels dissatisfied. A satisfying and realistic ordering of human affairs is one in which social structures reflect the objective equality of human beings. Once factual rather than formal standards of objective reason are identified we can recognize that the achievement of a human environment requires moving beyond a morality of want to a morality of needs and a conception of human nature defined not by individual life-plans, but by the social relationships and structure of needs within which such plans are made.
>
> (Simpson, 1979: 468)

Summary

The concept of RfP propounded in this chapter is Kantian as it uses morality to argue that RfP serves the interests of both the employee and the employer in construction. RfP has both individualistic and communitarian implications that underpin the nature of strategic and operational work in construction. Kant's three formulations of the fundamental principle of ethics have positive contributions to make in an industry where injury and fatalities go hand-in-hand with an increasing quest for profit. This RfP's insistence on moral motive requires a reorientation of management in the construction industry regarding the traditional focus on the bottom line. Rather than a narrow focus on profits and the proverbial 'business case', this RfP requires management to provide meaningful work for the employed in a safe and healthy workplace where contractors, subcontractors, and suppliers are not marginalised through power relations.

This chapter argues for respect as being the appropriate action required to enhance the health, safety and well-being of people in construction site work because respect helps to make moral sense of the notions of 'brother's keeper', and when permissible, 'zero harm' in the industry. The original RfP that is based on morality can be used to improve conditions for people working in construction. The anchor is that individuals in construction possess a dignity that is priceless and is to be respected as an end, which someone else cannot substitute. As much as the dignity of humanity is recognised, there is a moral obligation to respect oneself and others. The chapter has argued that construction enterprises have a duty to avoid violations of the rule of law, use of coercion and deception, unsafe working conditions, poor subsistence wages, and the abuse of the dignity of people in construction. Moreover, if the industry does continue to allow these problems, it will continue to disrespect workers regarding morality imperatives, which should not be undermined by economic considerations.

References

Abdelhamid, T.S. & Everett, J.G. (2000), Identifying root causes of construction accidents, *Journal of Construction Engineering and Management*, Vol. 126(1), pp. 52–60.

Adsul, B.B., Laad, P.S., Howal, P.V. & Chaturvedi, R.M. (2011), Health problems among migrant construction workers: A unique public–private partnership project, *Indian Journal of Occupational and Environmental Medicine*, Vol. 15(1), p. 29.

Aini, M.S., Fakhru'l-Razi A., Daud M., Adam, N.M. & Abdul Kadir, R. (2005), Analysis of royal inquiry report on the collapse of a building in Kuala Lumpur: Implications for developing countries, *Disaster Prevention and Management*, Vol. 14(1), pp. 55–79.

Anker, R., Chernyshev, I., Egger, P., Mehran, F. & Ritter, J.A. (2003), Measuring decent work with statistical indicators, *International Labour Review*, Vol. 142(2), pp. 147–178.

Arnold, D.G. & Bowie, N.E. (2003), Sweatshops and respect for persons, *Business Ethics Quarterly*, 13(2), pp. 221–242.

Auer, P. (2007), *Security in labour markets: Combining flexibility with security for decent work*, International Labour Office, Geneva.

Beauchamp, T. (2008), The principle of beneficence in applied ethics. In: E.N. Zalta (ed.), *The Stanford encyclopedia of philosophy* [Online]. Available at: http://plato.stanford.edu/archives/win2013/entries/principle-beneficence/ Date of access: 29 July 2016.

Bowie, N. (2000), A Kantian theory of leadership, *Leadership & Organization Development Journal*, Vol. 21(4), pp.185–193.

Bowie, N.E. (1998), A Kantian theory of meaningful work, *Journal of Business Ethics*, Vol. 17(9), pp. 1083–1092.

Bowie, N.E. (2002), A Kantian approach to business ethics, *A companion to business ethics*, pp. 3–16.

Buss, S. (1999), Respect for persons, *Canadian Journal of Philosophy*, Vol. 29(4), pp. 517–550.

Chi, S., Han, S. & Kim, D.Y. (2012), Relationship between unsafe working conditions and workers' behavior and impact of working conditions on injury severity in US construction industry, *Journal of Construction Engineering and Management*, Vol. 139(7), pp. 826–838.

Choudhry, R.M., Fang, D. & Mohamed, S. (2007), The nature of safety culture: A survey of the state-of-the-art, *Safety Science*, Vol. 45(10), pp. 993–1012.

Cooper, M. (2000), Towards a model of safety culture, *Safety Science*, Vol. 36(2), pp. 111–136.

Cranor, C. (1975), Toward a theory of respect for persons, *American Philosophical Quarterly*, Vol. 12(4), pp. 309–319.

Cranor, C.F. (1983), On respecting human beings as persons, *The Journal of Value Inquiry*, Vol. 17(2), pp. 103–117.

Darwall, S.L. (1977), Two kinds of respect, *Ethics*, Vol. 88(1), pp. 36–49.

De George, R.T. (2011), *Business ethics* (7th edn), Pearson Education, Upper Saddle River, NJ.

Dedobbeleer, N. & Béland, F. (1991), A safety climate measure for construction sites, *Journal of Safety Research*, Vol. 22(2), pp. 97–103.

Dekker, S.W. (2014), The bureaucratization of safety, *Safety Science*, Vol. 70, pp. 348–357.

Dekker, S.W., Long, R. & Wybo, J.-L. (2015), Zero vision and a Western salvation narrative, *Safety Science*, Vol. 88(2), pp. 219–223.

Emuze, F. & Khetheng, A. (2016), Workers' safety on construction sites: Use of PPEs in Lesotho. *Proceedings of the 9th Construction Industry Development Board (CIDB) Postgraduate Conference*, 1–4 February, 2016, University of Cape Town, Cape Town, pp. 220–228

Emuze, F., Van Eeden, L. & Geminiani, F. (2015), Causes and effects of building collapse: A case study in South Africa. *Proceedings of the Construction Industries Board (CIB) W099 International Health and Safety Conference*, 09–11 September 2015, Belfast, pp. 407–416.

Emuze, F., Van Eeden, L. & Geminiani, F. (2016). Regulatory factors contributing to building collapse in South Africa: A case study. *Proceedings of World Building Congress 2016*, 30 May–3 June 2016, Tampere, Finland, pp. 236–246.

Forbes, L.H. & Ahmed, S.M. (2010), *Modern construction: Lean project delivery and integrated practices*, CRC Press, Boca Raton.

Gaita, R. (2004), *Good and evil: An absolute conception*, Routledge, London.

Ghaemi, H. (2006), Building towers, cheating workers: Exploitation of migrant construction workers in the United Arab Emirates, *Human Rights Watch* November 2006, p.18.

Ghai, D. (2003), Decent work: Concept and indicators, *International Labour Review*, pp.113–142.

Glendon, A.I. & Stanton, N.A. (2000), Perspectives on safety culture, *Safety Science*, Vol. 34(1), pp.193–214.

Gray, M. & Stofberg, J. (2000), Respect for persons, *Australian Social Work*, Vol. 53(3), pp. 55–61.

Green, L. (2010, Two worries about respect for persons, *Ethics*, Vol. 120(2), pp. 212–231.

Guldenmund, F.W. (2000), The nature of safety culture: A review of theory and research, *Safety Science*, Vol. 34(1), pp. 215–257.

Hafey, R.B. (2015), *Lean safety Gemba walks: A methodology for workforce engagement and culture change*, CRC Press, Boca Raton.

Hudson, S.D. (1980), The nature of respect, *Social Theory and Practice*, Vol. 6(1), pp. 69–90.

Johnson, O.A. (1974), The Kantian interpretation, *Ethics*, Vol. 85(1), pp. 58–66.

Kant, I. (1949), *Critique of practical reason, and other writings in moral philosophy*, University of Chicago Press, Chicago.

Kant, I. (1974), *On the old saw: That may be right in theory but it won't work in practice*, University of Pennsylvania Press, Philadelphia.

Kant, I. (1993), *Grounding for the metaphysics of morals: On a supposed right to lie because of philanthropic concerns*, Hackett Publishing, Cambridge, MA.

Kant, I. (2012), *Fundamental principles of the metaphysics of morals*, Courier Corporation, North Chelmsford, MA.

Kant, I. (2013), *An answer to the question:'What is enlightenment?'* Penguin Books, London.

Kant, I. & Friedman, M. (2004), *Kant: Metaphysical foundations of natural science*, Cambridge University Press, Cambridge.

Kant, I. & Guyer, P. (1998), *Critique of pure reason*, Cambridge University Press, Cambridge.

Kuruvila, M., Dubey, S. & Gahalaut, P. (2006), Pattern of skin diseases among migrant construction workers in Mangalore, *Indian Journal of Dermatology, Venereology, and Leprology*, Vol. 72(2), pp. 129–132.

Levy, M. & Salvadori, M. (2002), *Why buildings fall down: How structures fail*, W.W. Norton, New York.

Lingard, H. & Lin, J. (2004), Career, family and work environment determinants of organizational commitment among women in the Australian construction industry, *Construction Management and Economics*, Vol. 22(4), pp. 409–420.

Lingard, H. & Rowlinson, S. (1998), Behavior-based safety management in Hong Kong's construction industry, *Journal of Safety Research*, Vol. 28(4), pp. 243–256.

Louden, R.B. (1986), Kant's virtue ethics, *Philosophy*, Vol. 61(238), pp. 473–489.

Lysaught, M.T. (2004), Respect: Or, how respect for persons became respect for autonomy, *Journal of Medicine and Philosophy*, Vol. 29(6), pp. 665–680.

Melé, D. (2009), Integrating personalism into virtue-based business ethics: The personalist and the common good principles, *Journal of Business Ethics*, Vol. 88(1), pp. 227–244.

Merrill, J.C. (2011), Overview: Theoretical foundations for media ethics. In: A.D. Gordon, J.M. Kittross, J.C.C. Merrill, W. Babcock & M. Dorsher (eds), *Controversies in media ethics* (3rd edn), pp. 3–32.

Modarelli, M. (2006), A Kantian approach to the dilemma of part-time faculty, *Changing English*, Vol. 13(2), pp. 241–252.

Mohamed, S. (2002), Safety climate in construction site environments, *Journal of Construction Engineering and Management*, Vol. 128(5), pp. 375–384.

Murray, M., Chan, P. & Tookey, J. (2001), Respect for people: Looking at KPI's through 'younger eyes'! In: A. Akintoye (ed.). *Proceedings of the 17th Annual Architectural Computer Services (ARCOM) Conference*, 5–7 September 2001, Salford, UK. Association of Researchers in Construction Management, 1, pp. 671–680.

Nash, L.L. (1990), *Good intentions aside: A manager's guide to resolving ethical problems*, Harvard Business Press, Cambridge, MA.

Ness, K. (2010), The discourse of 'Respect for People' in UK construction, *Construction Management and Economics*, Vol. 28(5), pp. 481–493.

Ness, K. & Green, S. (2008), Respect for people in the enterprise culture? Paper presented at the *24th Annual Architectural Computer Services (ARCOM) Conference*, September 2008, Cardiff.

O'Neill, O. (1989), *Constructions of reason: Explorations of Kant's practical philosophy*, Cambridge University Press, Cambridge.

O'Neill, O. (1993), *Kantean ethics*. [Online]. Available at: https://scholarblogs.emory.edu/millsonph115/2014/10/20/oneills-a-simplified-account-of-kantian-ethics/ Date of access: 29 July 2016.

Parekh, B.C. (2002), *Rethinking multiculturalism: Cultural diversity and political theory*, Harvard University Press, Cambridge, MA.

Parker, D., Lawrie, M. & Hudson, P. (2006), A framework for understanding the development of organisational safety culture, *Safety Science*, Vol. 44(6), pp. 551–562.

Philips, P. (2003), Dual worlds: The two growth paths in US construction, *Building chaos: An international comparison of deregulation in the construction industry*, pp. 161–187.

Place, K.R. (2010), A qualitative examination of public relations practitioner ethical decision making and the deontological theory of ethical issues management, *Journal of Mass Media Ethics*, Vol. 25(3), pp. 226–245.

Posthuma, A. (2010), Beyond 'regulatory enclaves': Challenges and opportunities to promote decent work in global production networks. In: A. Posthuma & D. Nathan, D. (eds), *Labour in global production networks in India*, Oxford University Press, New York, pp. 57–80.

Qizilbash, M. (1996), Capabilities, well-being and human development: A survey, *The Journal of Development Studies*, Vol. 33(2), pp.143–162.

Respect for People Working Group. (2000), *A commitment to people: "Our Biggest Asset". A report from the Movement for Innovation's working group on Respect for People*, Rethinking Construction, London.

Respect for People Working Group. (2004), *Respect for people: A framework for action*, Constructing Excellence, London.

Sample, R.J. (2003), *Exploitation: What it is and why it's wrong*, Rowman & Littlefield, Lanham, Maryland.

Sherratt, F. (2014), Exploring 'Zero target'safety programmes in the UK construction industry, *Construction Management and Economics*, Vol. 32(7-8), pp. 737–748.

Simpson, E. (1979), Objective reason and respect for persons, *The Monist*, Vol. 62(4), pp. 457–469.

Skrivankova, K. (2010), *Between decent work and forced labour: Examining the continuum of exploitation*, Joseph Rowntree Foundation, New Earswick, York.

Tam, C., Zeng, S. & Deng, Z. (2004), Identifying elements of poor construction safety management in China, *Safety Science*, Vol. 42(7), pp. 569–586.

Teuber, A. (1983), Kant's respect for persons, *Political Theory*, Vol. 11(3), pp. 369–392.

Wawrytko, S.A. (1982), Confucius and Kant: The ethics of respect, *Philosophy East and West*, Vol. 32(3), pp. 237–257.

Westphal, K.R. (1995), How 'Full' is Kant's categorical imperative? *Jahrbuch für Recht und Ethik/Annual Review of Law and Ethics*, Vol. 3, pp. 465–509.

Williams, M.C. (1992), Reason and realpolitik: Kant's "Critique of International Politics", *Canadian Journal of Political Science*, Vol. 25(1), pp. 99–120.

Young, S. (2014), From zero to hero. A case study of industrial injury reduction: New Zealand Aluminium Smelters Limited, *Safety Science*, Vol. 64, pp. 99–108.

Zwetsloot, G.I., Aaltonen, M., Wybo, J.-L., Saari, J., Kines, P. & De Beeck, R.O. (2013), The case for research into the zero accident vision, *Safety Science*, Vol. 58, pp. 41–48.

Index

work–family conflict and work–family culture 172–86

work–life balance 173–4, 184

working conditions in construction 8, 250–4

Workplace Employment Relations Survey (WERS) 134–7

World Cup construction projects 12, 214

Worth, H. 63

Wright, E. 235

Young, S. 231

'zero harm' concept 226–7, 230–9, 254; criticisms of 230–2; philosophical approach to 237–8; practical application of 235–8; reasons for adoption of 234–5; and safety culture 232–4

Zero Harm at Work Leadership Program (ZHWLP) 232–3

Zou, P.X. 236–7

Zwetsloot, G.I. 231